Handling laboratory microorganisms

Handling laboratory microorganisms

CHARLES PENN

OPEN UNIVERSITY PRESS
Milton Keynes · Philadelphia

Open University Press
Celtic Court
22 Ballmoor
Buckingham MK18 1XW

and
1900 Frost Road, Suite 101
Bristol, PA 19007, USA

First Published 1991

British Library Cataloguing in Publication Data

Penn, Charles
Handling laboratory microorganisms.
1. Microbiology. Laboratory techniques
I. Title
576.028

ISBN 0-335-09204-7
ISBN 0-335-09203-9 (pbk)

Library of Congress Cataloging-in-Publication Data

Penn, C. W.
Handling laboratory microorganisms / Charles Penn.
p. cm.
Includes bibliographical references and index.
ISBN 0-335-09203-9 (pb) ISBN 0-335-09204-7 (hb)
1. Microbiology – Technique. I. Title.
QR65.P39 1990
576'.078 – dc20 90–7785 CIP

Typeset by Vision Typesetting, Manchester
Printed in Great Britain by Redwood Press Limited

Contents

Preface

Standard texts of microbiology, whether elementary or comprehensive, usually omit practical advice, and it is perhaps assumed that practical training can only be acquired by word of mouth in the laboratory. For those encountering laboratory microbiology for the first time, whether as undergraduates or as practising scientists who need to exploit the powerful technologies of genetic manipulation and biotechnology, this book should provide a source of basic information either to back up laboratory classes or to prime the self-taught.

In a book of this size it is not practicable to provide comprehensive coverage of methodology. Other sources cited give detailed information on, for example, a wide range of media formulations, analytical procedures and systematic bacteriology. The intention is rather to provide an elementary description of the practicalities of handling microorganisms, with a summary of the relevant theoretical materials. This should enable beginners in the laboratory to work at the bench in a rational way, based on an appreciaton of some of the reasons behind the apparently ritual procedures which are commonly taught.

CHAPTER 1

Introduction – the significance of microbial sizes and numbers

Microorganisms are unique among the great diversity of life forms we know. Their properties of size and number dictate totally different concepts and practices of experimentation from those we adopt in studying 'macro'-organisms. Furthermore, microorganisms provide unique opportunities for investigation of biological phenomena and processes. It is no accident that they have provided fundamental knowledge of the genetic material, and the tools for working out the genetic code. Current knowledge of molecular biology, genetic engineering and biotechnology would not have developed without the exploitation of all the varied opportunities for experimental approach offered by microorganisms.

Size and number are interdependent: vast, unimaginable populations of microorganisms can be handled routinely in the laboratory only because the individual units are so small. The experimental microbiologist must have a firm idea of the sizes and dimensions, and units of measurement, involved in microbiology. The most important dimension is the *micron*: one millionth of a metre, or one thousandth of a millimetre, designated 1 μm. Incidentally, verbal expressions of these fractions, and likewise of very large numbers, are clumsy and liable to misinterpretation, and a routine mental process the student should master is their manipulation in logarithmic form. One millionth of a metre should be expressed as 10^{-6} m, and a thousandth of a millimetre as 10^{-3} mm. The use of logarithms is vital in handling serial dilutions and titres as well as for very large numbers. I shall return to these conventions in Chapter 9.

The micron is useful because it is a convenient unit of measurement for many microorganisms. It is also quite close to the limit of resolution of the light microscope. The 'prototype' microorganism, to which I and other microbiologists will return repeatedly, is *Escherichia coli* (named rather clumsily, like many bacterial genera, after the first man believed to have described it, Theodor Escherich). *Escherichia coli* usually has dimensions of about 1.5 × 2.5 μm. Why 'usually'? Because many properties of microorganisms are variable, as we shall see later. To illustrate microbial size effectively in terms familiar in everyday life is not easy, but I have tried to so by the comparisons in Table 1.

So, the dimensions of many bacteria (the primary subject of this book, although I shall refer to unicellular fungi occasionally, but not viruses) are of the order of microns. Small cocci are approximately spherical, with a diameter of about 1 μm. Consider an idealized microorganism which is a cube with sides of 1 μm – i.e. of the same order of volume as many bacterial cells. It is instructive to think about the number of such cubes

Table 1 Size comparisons: Just how small are microorganisms? The following examples may help to give some appreciation of the scale of microbial life.

Object	*Size (approximate diameter)*	*Number relationship (approximately) with everyday objects*
Pinhead	2 mm	250 would cover a postage stamp. 15 000 would cover this page. A million would cover most of the floor of an average bedroom.
Hair	0.02 mm	A bundle of 10 000 could be attached to a pinhead. One hundred and fifty million, on end, would cover this page.
Bacterium (1 μm diameter coccus)		250 could cover the cross-section of a hair. More than a million would cover a pinhead – as pinheads would cover your floor.

which could be contained within volumes we are familiar with. Since there are 10^3 μm in 1 mm, there will be $10^3 \times 10^3 \times 10^3 = 10^9$ μm^3 in 1 mm^3. There are also $10 \times 10 \times 10 = 10^3$ mm^3 in 1 cm^3. The latter is a familiar volume, and it will thus contain $10^9 \times 10^3 = 10^{12}$ μm^3. That is, one million million, or 1 000 000 000 000 volumes each approximating to that of a small bacterial cell can be packed into 1 cm^3 (or 'cubic centimetre', or 'millilitre', whichever was the current convention when we were taught these things!). The human population of the world is currently about 5×10^9. It is a reasonable guess that the total number of humans there have ever been is of the order of 10^{12}. We could not even conceive of getting even 10^6 (one million) people (or mice, or any other 'macro'-organism familiar to biologists) together in one place, packed tightly together. But the population of 10^{12} bacteria indicated above, in a volume of 1 ml (or slightly more, since cells are not easily packed together so tightly), can be obtained routinely in a microbiology laboratory. *Escherchia coli*, growing in optimal conditions in a reasonably rich medium and starting from a concentration of 1 cell ml^{-1}, will divide about every 15–20 min. Its numbers therefore increase by about tenfold every hour – 'exponential' growth – to which we shall return in Chapter 7. It is useful to remember that a tenfold increase, easy to handle in the convention whereby powers of 10 or logarithms to the base $_{10}$ ($\log_{10}$) are usually used in analysing data containing very large or very small numbers, represents between three (eightfold) and four (sixteenfold) doublings in number, or 'generations' (Fig. 1). It will therefore take about 9 hours (i.e. about nine tenfold increases or 9 'logs' increase in the usual jargon) for 1 cell ml^{-1} to become 10^9 ml^{-1}. In ordinary media, without careful control of factors such as build-up of acidic metabolites, the culture will stop growing at about this density. Given 1 litre of such a culture, easy to produce in the laboratory, we could have produced in 9 hours, or about 10 hours if we started from one cell, a population of bacteria approximating the total number of humans who have ever lived.

There are many important implications in these extraordinary concepts of size and number. First, size itself affects a very significant physical parameter – the ratio of surface area to volume. If we consider the 1-μm cube again, it will have a total surface area of 6 μm^2. The 10^9 1-μm cubes which occupy 1 mm^3 therefore have a total surface

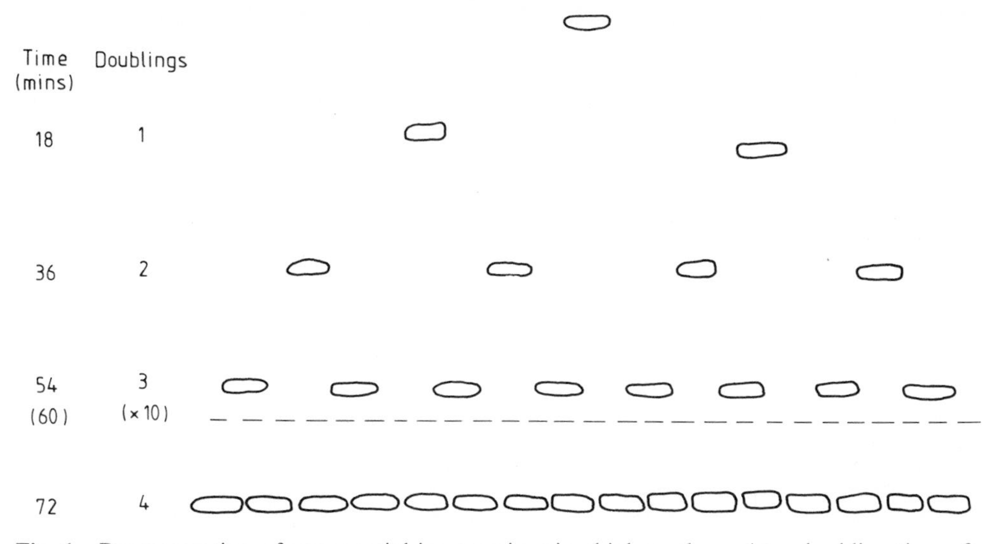

Fig. 1 Representation of exponential increase in microbial numbers. At a doubling time of about 18 min, typical of *E. coli* growing at its optimum rate in complex medium, a tenfold increase in numbers occurs in about 1 h (dashed line).

area of 6×10^9 μm^2, or 6×10^3 mm^2. On the other hand, a single 1-mm cube will have a surface area of only 6 mm^2. So for a total volume of 1 mm^3, our two examples have a one thousandfold difference in total surface area: the ratio of surface area to volume increases dramatically as size decreases. This is a major factor determining, for example, the capacity of microorganisms for rapid growth: they have a relatively enormous surface area for rapid transport of nutrients or metabolites into or out of cells. In larger organisms these are likely to be rate-limiting processes for growth.

Second, size dictates an unexpected disadvantage in some respects in the study of microorganisms: we are very seldom able to examine the properties or behaviour of individual cells. Mostly we observe populations, and phenomena we can record are usually the average for a population. We cannot usually observe the growth and division of a single cell, and dissect the processes going on within it in any detail. Nevertheless, we know a great deal about these processes from the study of populations. This brings us to another fundamentally important concept: how uniform are the members of microbial populations? The numbers in populations being studied become very important in answering this question.

The properties of each cell are ultimately determined by the information encoded in its genome. Theoretically, a population which has arisen from a single cell (or a single individual parent) by repeated non-sexual division may be considered a 'clonal' population: every individual should be genetically identical. (The ability of micro-organisms to produce large clonal populations rapidly from single cells is in fact the basis of gene cloning: each member of a genetically manipulated clone has identical copies of a manipulated gene. Without the tools of microbiology, gene cloning certainly would not have developed.) In 'macrobiology', the uniformity of clonal populations would usually be absolute, since we do not usually encounter very large

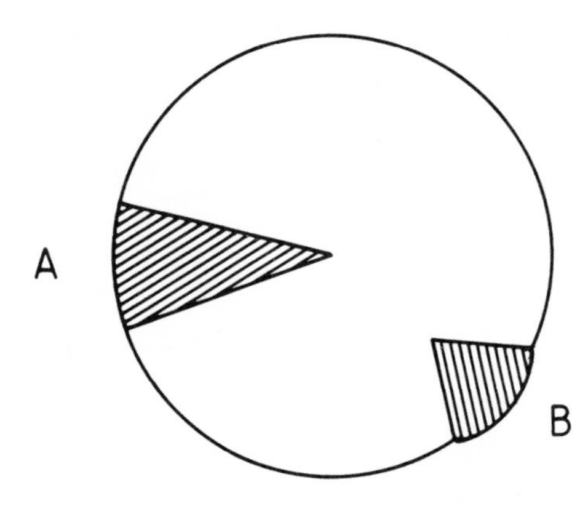

Fig. 2 Typical effects on the morphology of a colony resulting from mutations which affect the colour or texture of the colony or growth rate of the cells. At A there is only a change of colour or texture, which occurred very early in the growth of the colony, hence the sector arises near the centre. At B, the mutation arose late during colony growth, and resulted also in more rapid growth of the cells so that the sector is bulging away from the colony at its periphery.

clonal populations. In microbiology, however, the situation is very different. It is commonly accepted in biology that most genes mutate spontaneously on average after about 10^6 to 10^8 cycles of replication. Alternatively, we can express this as a 10^{-6} to 10^{-8} (i.e. 1 in 10^6 to 1 in 10^8) probability that any given cell will contain a mutation in any particular gene, or, we may say that the latter is the probability that such a mutation will occur at any given cell division. Yet another way to express this is that such a mutation will be found once in every 10^6 to 10^8 cells of a clonal population. Thus in the easily obtained 10^{12} cells discussed above, or even in 1 ml of an overnight culture containing 10^9 cells, there will almost certainly be a number of cells which carry mutations in any gene we may wish to examine. This is a situation quite unlike those we encounter in other areas of biology, and is the basic reason why microbial species have been so central to the development of molecular biology and molecular genetics.

There are important practical consequences of this 'mutability' of microorganisms. Some routine laboratory manipulations select clonal populations, and in the absence of obvious markers that the characters we are interested in have not mutated, it is quite possible accidentally to pick an atypical clone. This discussion so far has focussed on the majority of reasonably stable genes which have been well studied in the past. We are now realizing increasingly that in many microbial species there are mechanisms for more rapid genetic change. For the very reasons outlined above, organisms or characters which exhibit high frequencies of spontaneous mutation (in some cases as high as 10^{-2} to 10^{-3} per cell division) are difficult or inconvenient to handle experimentally. Microbiologists have probably tended to select relatively stable organisms or genetic systems for many types of experimental study. In the case of genetic characters which change at high frequency, this sometimes leads to easily seen alterations in colony morphology and the origin and development of mutated clones can be observed *in situ* in clonal populations, i.e. colonies derived from single cells. An example is shown in Fig. 2. The alert microbiologist may notice such phenomena quite frequently, and often they will be unexplained; this is a science which will constantly provide stimuli for curiosity-driven investigations arising from such chance observations!

CHAPTER 2

Equipment and materials

A certain amount of elementary microbiology can be undertaken with surprisingly little equipment, especially since the development of pre-packaged and sterilized disposable plasticware. It will however be useful to outline the range of equipment and the type of laboratory needed for full-scale research work, with comments where possible on ways to cut corners for work on a limited budget.

The laboratory

Like all experimental sciences, microbiology is potentially hazardous, and there are particular biological hazards associated with some aspects. These will be dealt with in more detail in Chapter 6. In choosing a laboratory, however, there are one or two basic rules which are important. The room should not be a through route for movement of personnel, for two reasons: (1) if potentially hazardous cultures are being handled, risk of infection is increased; and (2) vigorous movement of any kind is likely to create or stir up dust particles, the main source of contamination of cultures. For the same reason it should be possible to close doors and windows, particularly while work is in progress but also at other times so that vigorous air currents can be avoided and dust can settle out of the air. To avoid accumulation of dust, the room should be kept free of clutter, and at intervals all horizontal surfaces should be cleaned. Cupboards and drawers should be used as far as possible for storage rather than open shelves. Lighting should be bright and diffuse, but a directed light source such as an angle-poise lamp should also be available. Bench surfaces should be impervious and easily cleaned, preferably made of high-quality synthetic laminate. Benches should be designed to avoid nooks and crannies around shelving and service outlets, for example by use of a step of several inches at the back of the bench. Numerous power points, at least four for every metre of bench, are needed in a research environment, fewer in a teaching laboratory. Gas, water and drainage should be available at every workstation, and a sink should be readily accessible to every worker. Ideally a vacuum line should be accessible at every workstation, but if mains water pressure is good, water-operated suction devices are useful for many purposes.

Thought should also be given to the 'domestic' arrangements of personnel. Writing and reading should be physically separate from benches, and ideally in a separate room with both direct and independent access. It should not be necessary for outdoor clothing, lunch or shopping to be stored in the laboratory. A separate room for eating, drinking, smoking and seeing visitors is essential if these activities are to be excluded from the laboratory successfully. There should be a separate sink near the exit for hand washing, preferably with lever-operated taps. A suitable fire extinguisher, for example a carbon dioxide model compatible with electrical hazards, should also be near the door and be prominent.

Equipment

STERILIZATION

The most basic need is a facility for sterilization, and in practice this means an autoclave. For small-scale, elementary work, such as the most basic teaching, this need can however be avoided by purchase of pre-prepared sterile media ready for use, and a safe means of disposal of small amounts of culture might be, for example, incineration or carefully controlled treatment with heat or chemicals. An ordinary domestic pressure cooker could also be used for small-scale sterilization and disposal provided it is used properly (see Chapter 5). It must be remembered that all cultures should be treated as potentially hazardous: even if the organism grown is known to be harmless it may, unknown to the operator, be contaminated with a potential pathogen originating from a human source. Furthermore, organisms which are harmless in the small numbers routinely encountered in the environment may become hazardous when grown to high numbers or concentrations in the laboratory. Nevertheless, with understanding and common sense, which I hope to convey in this book, no-one with a basic grounding in scientific principles should be deterred from doing microbiological work. There is no mystique which cannot be rationally explained, and no procedure which need be alarming if performed with full understanding.

For any serious, sustained level of laboratory microbiology then, an *autoclave* is essential. In molecular biology laboratories where DNA or RNA are routinely handled, autoclaving is used not merely to sterilize but also to destroy nucleases which are ubiquitous and often very stable proteins. For small-scale work, the autoclave may be quite a small, benchtop model powered by gas or electricity and costing only a few hundred pounds. For larger scale work, a floor-mounted autoclave of substantial cost may be necessary. With larger models steam production may be a problem and suitable ventilation will be needed, so a separate room will be necessary. Such a room may usefully be a washing-up room and may be used for preparation of sterile supplies as well.

One more item of sterilizing equipment is an *oven*, useful for some glassware such as pipettes which have to be kept dry. An oven with fan, timer and accurate temperature control will cost several hundred pounds.

Incidentally, not all sterilizing has to be done by heat (see Chapter 5). For example, small items such as dissecting instruments or glass rods may be sterilized effectively by dipping in 70 per cent (v/v) ethanol in water, shaking and flaming to burn off the remaining alcohol. For many purposes absolute sterility is not strictly necessary, and essentially 'sterile' equipment can be obtained by boiling, for example. Much depends on the application: rich liquid culture media, to be incubated for long periods with small numbers of slow growing organisms, are ideal for multiplication of small numbers of contaminants and glassware must be strictly and absolutely sterile. On the other hand, in petri dishes for solid media where occasional contaminants will be highly visible and localized, a low level of contamination may be acceptable. I heard of one impoverished laboratory where disposable plastic petri dishes were routinely scraped clean, dipped in hypochlorite, rinsed in tap water and dried at 60°C: they proved to be essentially sterile and re-usable, but I would not recommend this except in extreme circumstances!

DISPOSAL OF CONTAMINATED MATERIAL

All culture of microorganisms produces potentially hazardous material for the reasons given above. It is therefore essential to have effective 'discard' procedures. An understanding of microbial killing agents and mechanisms is needed (see Chapter 6). For routine use, the most effective agent is heat. For immediate heating of, for example, wire loops or glassware necks, the traditional bunsen burner is efficient, cheap and reliable. Ideally it should have a pilot jet so that it can be turned down when not in use, avoiding fumes and heat production (a problem in a teaching lab with many running at once), and lessening the risk that hair or clothing will be burnt accidentally. The pilot flame must however be large enough to be visible.

The second main requirement for discard procedures is a supply of impervious, heat-proof containers for collection of discarded cultures and vessels. Discard containers should be covered but not sealed as they will normally be autoclaved with contents before disposal. Stainless-steel buckets with lids are convenient and cheap and will last almost indefinitely if protected from acids and hypochlorite. Ideally, disposable plastics and other material such as swabs should be discarded separately from glassware so that plastics do not melt and weld together small items of glassware. Autoclavable plastic bags can be used for subsequent ease of disposal of these items. Alternatively, they may be disposed of directly, for example by incineration, provided foolproof procedures for their safe collection and handling can be made. For discard into bags, stands with lids operated by pedals are useful, or bags may be taped to the front of the bench (Fig. 3).

Specialized containers are needed for discard of pipettes. Graduated glass pipettes are now used less as laboratories move towards miniaturized technology and disposables, but they remain essential for some purposes. Glass pasteur pipettes, if recycled, are still cheap and effective. It is best to discard graduated and pasteur

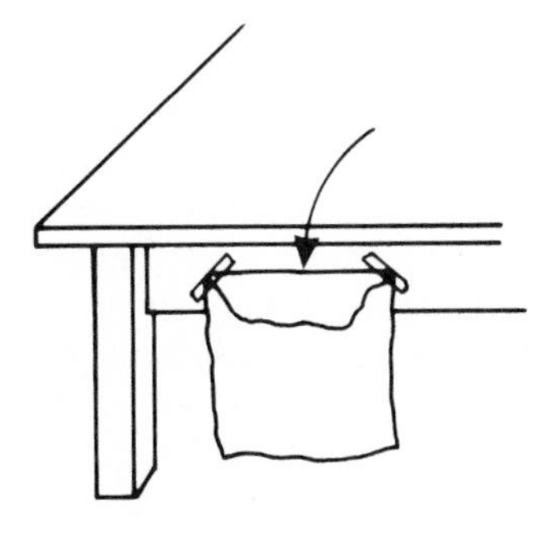

Fig. 3 A waste bag (autoclavable) taped to the side of the bench is convenient for the discarding of lightweight, dry waste such as small disposable plastic items.

pipettes separately, in each case into containers deep enough to immerse the pipette completely in disinfectant. The merits of various disinfectants are discussed in Chapter 6. Purpose-made containers in rubber or plastic can be bought.

A particularly hazardous category of waste is 'sharps' such as needles and blades. These should be discarded into specially made disposable containers which can be autoclaved or ideally incinerated without further handling. Broken glass should be collected separately from other rubbish and disposed of carefully, after autoclaving if it is contaminated.

INCUBATORS, WARM ROOMS AND SHAKERS

Facilities for incubation at controlled temperature are essential for scientific work. Even if organisms grow at room temperature this is normally variable and uncontrolled. For incubation at higher temperatures than ambient, a wide range of incubators is available. To obtain the most constant temperatures with frequent opening for access, considerable sophistication is needed. A water jacket will act as a heat reservoir, returning the temperature rapidly to its set point by heat transfer through its conductive metal lining of large surface area. Fan circulation for even temperature is no substitute if moving air will jeopardize sterility, e.g. of cultures in petri dishes. Incubator control systems may be complex, with overheat cut-out and alarm as well as an adjustable thermostat. Larger incubators will tend to have greater stability. Thus a good reliable facility will be an expensive item.

To maintain constant temperatures at or below ambient, additional complexity results from the need for cooling as well. For occasional use an effective substitute is to locate an incubator or water bath in a cold room. It should also be remembered that in some weather and especially if ventilation is poor and there is much equipment producing heat, ambient temperatures may become quite high. Fan-assisted incubators or shaker cabinets, in which the mechanism itself may produce heat, may then overheat even if set to run at several degrees above ambient (e.g. at 37°C when ambient temperatures are $>30°C$). Just because equipment functions well thoughout the winter, it will not necessarily do so in a heat wave!

A specialized, and even more expensive, item is the carbondioxide (CO_2) incubator (Fig. 4). It is normally humidified and programmed to maintain a level of about 5 per cent CO_2 in air, to provide efficient carbonate/bicarbonate buffering in tissue culture media. There is a fan to circulate the atmosphere so that its composition remains constant, a sensor (infra-red absorption or thermal conductivity) to monitor CO_2 level, valves to control injection of CO_2 from a cylinder, and a water reservoir of large surface area to maintain high humidity so that media exposed to the CO_2-enriched atmosphere (about 5 per cent) do not dry out.

Incubation in the gas phase is suitable for cultures of low thermal capacity (e.g. petri dishes) and/or for long periods of incubation (at least overnight) so that thermal equilibration occurs after a small proportion of the total incubation time. For short-term incubation, especially of liquids, water baths provide much more efficient heat exchange. Again, a wide variety is available, but beware of cheap equipment: they need to be reliable and durable, since they are often abused through being left on all the time even when not in use, run at too low a level, or not cleaned out for long periods. For most purposes, circulation of the water by a propeller is desirable, to provide an even temperature and rapid heat exchange. Shaker water baths are also available. To cut

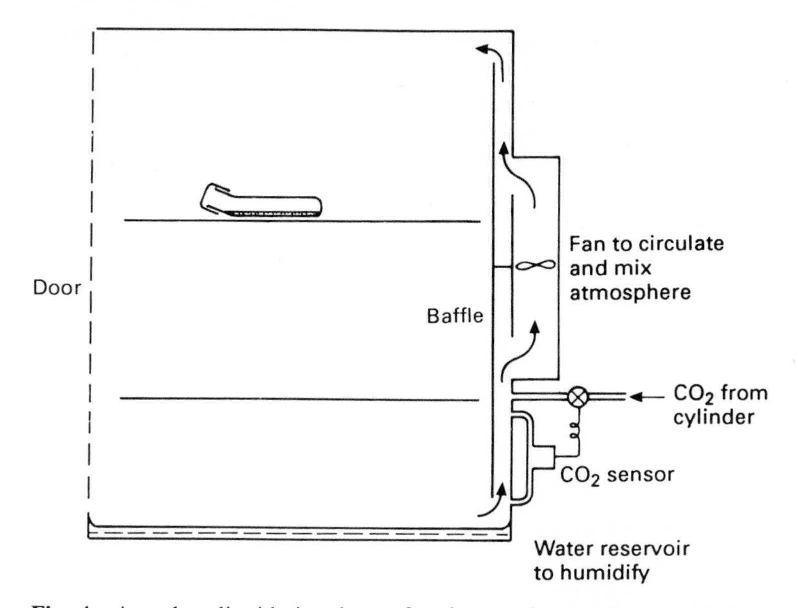

Fig. 4 A carbondioxide incubator for tissue-culture cells. In addition to the mechanism for addition to the atmosphere and its circulation of CO_2, the incubator would be heated by a conventional electric element, and may or may not have a water jacket.

down on evaporation, lids should be used when possible and it may also be feasible to float ping-pong balls on the surface, thus reducing heat loss as well. An asset for rapid warming of small volumes of liquid in tubes is the heating block (Fig. 5), a solid aluminium block drilled to take tubes, especially of the Eppendorf type (see Fig. 9 on p. 15), but also available for others. It is heated electrically and the high thermal conductivity ensures rapid and even heat distribution.

Shakers are essential for efficient liquid culture for many experiments – yields are often higher, and culture conditions more reproducible and homogeneous (see

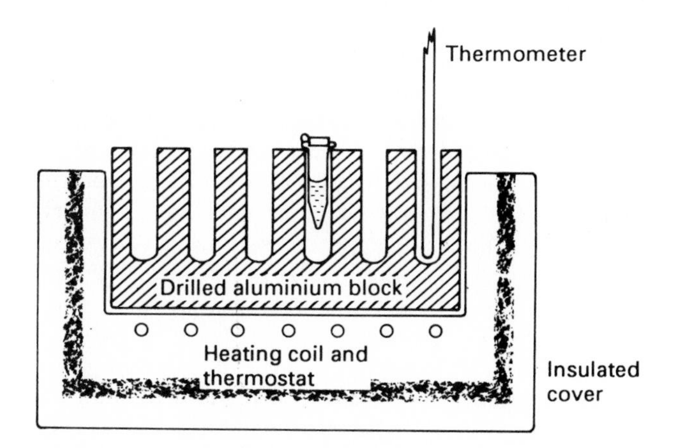

Fig. 5 A heating block, designed for incubation of Eppendorf or similar small tubes. Heat transfer is rapid, but there is not the inconvenience and mess of a water bath.

Chapter 8). Cabinet shakers with controlled temperature are ideal, but need to be very solidly constructed if of any size, and are expensive – several thousand pounds for a platform of about 0.3 m^2. Alternatively, orbital or reciprocating shaker platforms can be located in warm rooms, or small ones may be put inside incubators. Often inadequate is the range of clips available for holding conical flasks. For the large shakers at least a dozen should be provided for each of the smaller sizes of flask (25, 50, 100, 250, 500 ml), and half a dozen for 1 litre and 2 litre flasks. Another possibility for non-static cultures is the use of magnetic stirrers, with followers autoclaved in the culture container.

A *warm room* is useful for several purposes in a large laboratory, although again for reliability the heating and control equipment must be good quality and may be expensive to buy and maintain.

FRIDGES, FREEZERS AND COLD ROOMS

Ordinary domestic refrigerators (normally operated at 4°C) and freezers are quite good for laboratory use, although lightweight plastic fittings may sometimes break in heavy use. Factors of design and convenience apply just as in the home. Domestic freezers which on the lowest setting reach –20°C are adequate for short-term storage of cryo-protected (e.g. with glycerol) cultures of robust microorganisms, and for media, enzymes and sera, and many stock solutions, but not for cell lines or virus stocks. In busy labs where freezers will be opened frequently, chest rather than upright models may be less liable to frosting up, but they need more floor space and the contents are less accessible.

For long-term storage or for sensitive organisms such as viruses, or for cell lines, a –70°C freezer is necessary. This requires much mechanical sophistication and heavy-duty insulation and is very expensive – again, several thousand pounds. There is considerable heat production from these freezers, and special arrangements for ventilation or cooling in hot weather may be needed. An alternative is the *liquid nitrogen freezer* which has no moving parts and produces no heat, but it does require topping up weekly, and hence regular deliveries of liquid nitrogen and a storage vessel are required. Again, a substantial running cost is involved, and occasionally there may be accidental losses when someone forgets to top up the freezer. In other respects these freezers are extremely reliable. For large laboratories, walk-in cold rooms, at 4°C and –20°C, are convenient. However cold storage is subject to a type of 'Parkinson's law' – the material expands to fill the available space, and it is amazing how much turns out to be junk when the system breaks down and only the precious items can be stored elsewhere! This raises the problem of spare space and discipline – the first depends on the second. Spare space is essential both for defrosting, which must be done regularly to maintain the efficiency of the freezer, and for emergency use during equipment breakdown or maintenance.

STERILE BENCHES AND SAFETY CABINETS

Dedicated microbiology laboratories will require a sterile bench or sterile area for use when absolute sterility is vital, for example in handling long-term cultures of cell lines. The problems of sterility and aseptic technique are discussed in Chapter 5. The most effective form of sterile area is the laminar flow cabinet (Fig. 6), in which a fan pushes

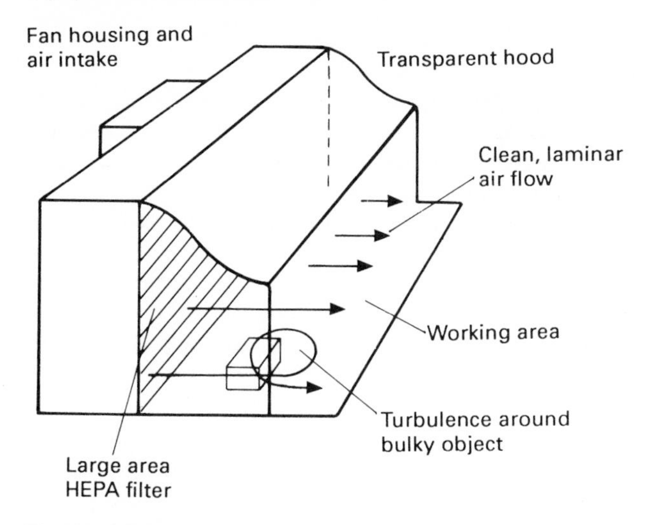

Fig. 6 A laminar flow hood for aseptic handling of cultures, materials or equipment. The filter, covering the whole area of the back wall, provides a laminar flow of sterile air without turbulence, but this may be disturbed by bulky objects and turbulence may then draw dirty air into the working area.

air through a HEPA (High Efficiency Particle Air) filter, which entirely fills the back of the working area, and which is capable of removing all particulate matter from the air down to a size which ensures sterility. The sterile air moves towards the operator in a laminar (i.e. non-turbulent) flow, so that manipulations upstream of the operator take place in a sterile atmosphere, provided there is no turbulence causing backflow of dirty air from the direction of the operator (Fig. 6). This type of facility is also used in manufacturing processes in which dust must be strictly excluded from delicate mechanisms. For maximum effectiveness the room air around the front of the cabinet must be still. When used properly for experiments in which sterility must be guaranteed, such cabinets are highly effective, but for much routine microbiology they are not essential. Their cost is moderately high – at least several hundred pounds even for the most basic unit. An effective alternative may be a small, uncluttered separate room in which dust and air movement can be kept to a minimum, used only for sterile procedures.

An exactly opposite requirement to the maintenance of sterility is containment of airborne microbiological hazards (see Chapter 6). *Safety cabinets* must not be confused with sterile benches! The air stream in safety cabinets is intended to carry airborne hazards away from the operator and into a filter so that clean air can be forced back into the laboratory or, for extra safety, ducted to the outside air. A moment's thought will show that a simple arrangement which is the reverse of a sterile bench (i.e. a class I safety cabinet) will be very likely to contaminate experiments. If the work requires sterility as well as safety, a class II cabinet, in which a downflow of sterile air at the front of the cabinet is combined with a backflow to a filter and exhaust away from the operator, may be used. Further details of the use of safety cabinets are given in Chapter 6. They will normally only be needed for dangerous pathogens, or for containment of certain categories of genetic manipulation work.

MICROSCOPES

The microscope is perhaps the most important instrument used by the microbiologist, although curiously it is often not used to maximum advantage. Microscopy is a science in itself and indeed to some extent an art: dedicated microscopists are interested in more than just a tool for routine use in a busy laboratory. Therefore, microscopes cover a very wide range of complexity and cost. Selection of a microscope will depend very much on the intended application.

First let us return to the question of microbial size. Particles of the 1 μm dimension referred to in Chapter 1 really dictate a minimum magnification of ×400 for routine observation; this can be obtained with a ×40 objective and ×10, not only for scanning a wider field but also for initial focussing – often very tricky for the inexperienced user. A ×100 objective is also necessary, for more detailed observation of bacterial morphology at a total of ×1000 with the ×10 eyepiece. Such an objective requires oil immersion, but although messy, its use is essential for examining stained smears of unidentified cultures. This operation requires brightfield optics – the cheapest and simplest to use. For many purposes other than identification, however, a system for examination of wet mounts is also very useful. The most common is *phase contrast*, which exploits differences in refractive index between specimen and diluent and hence allows instant visualization of unstained, wet-mounted bacterial cells. The system is however a substantial additional cost. Some saving will be made by limiting phase contrast to one or two objective lenses. An alternative to phase contrast is *dark ground*, again exploiting the higher refractive index of cells as against that of the suspending medium. It may be cheaper than phase contrast and is very effective for examining slender spirochaetes, a traditional use being the diagnosis of syphilis by microscopy! There is, however, more interference from dust particles and especially bubbles, and oil immersion between the slide and the condenser is necessary for high power – very messy! High-power objectives for phase contrast may also require a special aperture stop.

An additional major factor determining cost is the choice of monocular or binocular. For anything other than very occasional or transient use, a binocular instrument is strongly recommended: it substantially reduces eyestrain when used routinely.

More sophisticated requirements in microscopy are for photography and fluorescence microscopy. For *photography*, a three-way 'tube' (the central part of the optical system, between objective and eyepiece) is necessary so that the image can be directed into the camera, and of course a suitable camera which, since it does not require a sophisticated lens, is relatively cheap. *Fluorescence* facilities again increase the cost greatly. They need not necessarily include ultraviolet illumination – high intensity quartz–halogen–tungsten bulbs emit useful amounts of short wavelength visible light, and fluorescein, a commonly used fluorochrome, can be excited efficiently by blue light. The most useful illumination for fluorescence is epi-illumination, in which a half-silvered mirror is used, with a suitable filter combination, to enable both downward reflection through 90° of incident light onto the specimen from above, and upward transmission of emitted light to the eyepiece (Fig. 7). This avoids the need for complex condenser systems to handle the incident illumination, and also allows examination of opaque objects.

One or two specialized types of microscope should also be mentioned. An *inverted*

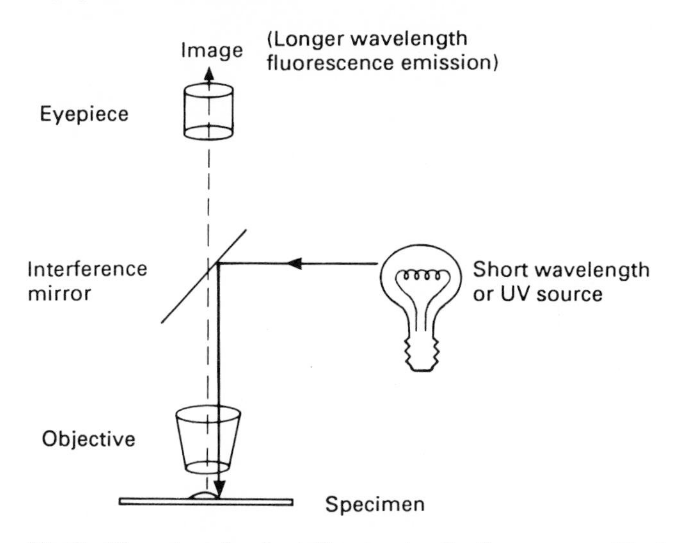

Fig. 7 The principle of epi-illumination for fluorescence. The interference mirror is the critical component: it selectively reflects short wavelength light (solid line), while transmitting to the eyepiece longer wavelength light (dashed line) emitted by the specimen.

microscope is almost essential for tissue culture work, and can be used to examine cells growing on the bottom of tissue culture vessels – hence they are examined from below (Fig. 8), since the distance to the specimen from above the container is too far for high magnification lenses. Routinely, such a microscope is used at quite low power, and a good quality routine instrument with phase contrast would cost about £1500. Another specialized microscope is the *stereoscopic dissecting microscope*, sometimes used for detailed examination of colony morphology: it has twin objectives as well as twin eyepieces. This has the disadvantage that observations cannot be recorded fully (i.e. three-dimensionally) by photography.

The suppliers of microscopes are very diverse, and their costs very variable. Good quality, low-cost instruments are available from the Far East, particularly Japan. A simple, brightfield-only instrument capable of image quality adequate for routine bacteriology may cost only a few hundred pounds, and phase contrast could be added without taking the price much above £1000. The refinements of, for example, wider and flat fields and achromatic optics add considerably to the cost and will only be available on more expensive instruments. Mechanical reliability and durability, perhaps adequate for a lifetime's use, also imply high cost. A top-range instrument with the capacity to add on a variety of photographic, fluorescence and other attachments, from a leading manufacturer such as Zeiss or Nikon, might cost five times as much as a cheap, basic brightfield instrument.

BALANCES

In any but the most basically equipped laboratory at least two balances are needed. For routine weighing of chemicals and media in relatively large quantities, a wide range of simple, reliable top pan electronic balances are now available for less than £100. Operation by push button greatly increases their speed of use and convenience. For

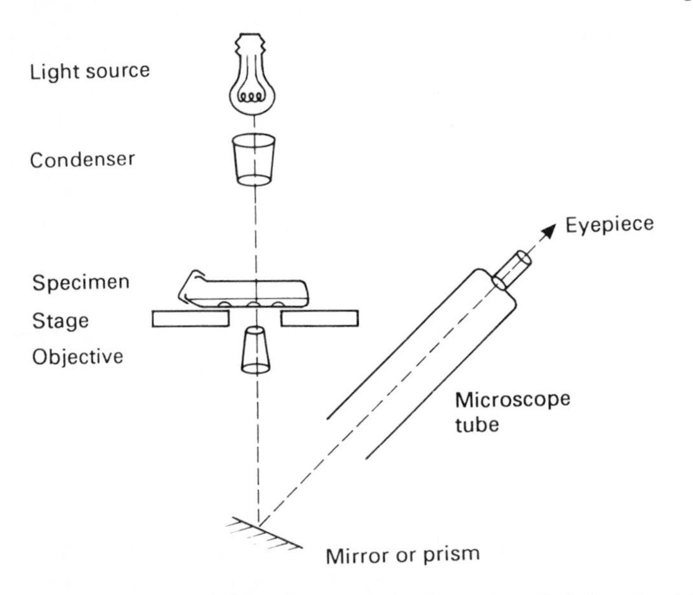

Fig. 8 The principle of the inverted microscope. To bring the objective sufficiently close to the object, located on the lower surface of a culture vessel, the objective is mounted below the specimen, and a mirror or prism reflects the light path upwards into the tube for examination. The light source and condenser are above the specimen.

more precise weighing of small amounts of precious reagents, a sophisticated and accurate fine balance weighing to 1 mg or less is also required – quite an expensive item at perhaps £1000.

pH METER

Every laboratory should have a pH meter, again perhaps with the exception of the most basic teaching laboratory. (Universal and narrow-range indicator papers can be surprisingly effective, particularly on a very small scale – 1 μl of liquid will produce a coloured spot on paper.) Again, electronic circuitry has lowered costs and improved reliability and convenience in the use of pH meters. Simple battery-operated 'dipstick' models with liquid crystal display are excellent for routine use checking media etc., and can be bought for well under £100. This compares with at least several hundred pounds for a conventional mains-powered benchtop model. The latter type usually has additional refinements such as a millivoltmeter option and temperature compensation which are seldom used in most microbiology laboratories. There is however the advantage that a variety of specialized electrodes can be used, for example miniature versions for small volumes.

Combination electrodes are commonly used with small volumes, and may be only a few millimetres in diameter. It is wise to use an electrode with a built-in guard to prevent accidental knocking of the glass membrane, which is extremely delicate – replacement electrodes are quite expensive.

CENTRIFUGES

The centrifuge is an essential item for the serious microbiologist. To pellet bacteria, accelerations of several thousand ***g*** are needed. A standard benchtop centrifuge, able to handle volumes of 10–50 ml, and providing speeds of 3000–5000 r.p.m., is usually the basic requirement. It will cost at least several hundred pounds, and should have basic safety features such as a lid interlock device to prevent accidental opening while it is running.

A recent development which in many applications supersedes the benchtop machine is the *microfuge*, which has become standard in molecular biology laboratories. This is a product of the trend towards miniaturization – others are microtitre systems, micropipettes and mini-electrophoresis tanks. Enormous savings in precious materials, reagents and time can result from their use. The microfuge is a simple but very powerful device – it relies on the fact that a given ***g*** force will pellet particles more quickly if they have a shorter distance to travel to the bottom of the tube. The tubes used have become standard, disposable items, after being developed largely for use in clinical chemistry: the so-called Eppendorf tube (Fig. 9). Microfuges normally provide speeds of 10 000–12 000 r.p.m., and can pellet bacteria in 5 min in such tubes, with ***g*** forces around 10 000–15 000. Because run times are short (due to the short tube length), heat generation is not a problem. In conventional centrifuges running at this speed, refrigeration would be necessary, due to the friction of rotor against air. A number of different microfuges are available, many costing less than £1000.

In research laboratories requiring facilities for larger scale work or higher ***g*** forces, there are four main categories of machine. A general-purpose refrigerated floor-standing centrifuge, capable of speeds up to about 20 000 r.p.m. and for volumes up to about 1.5 litres, generating ***g*** forces approaching 50 000 depending on rotor design and capacity, is probably the most generally useful and will cost several thousand pounds. Secondly, a larger capacity machine to handle up to 6 litres, but capable only of speeds up to about 6000 r.p.m., may be useful if large volumes of culture are to be processed. Again it will be refrigerated and of comparable price to the first. Thirdly, a free-standing ultracentrifuge will be needed for basic research programmes involving

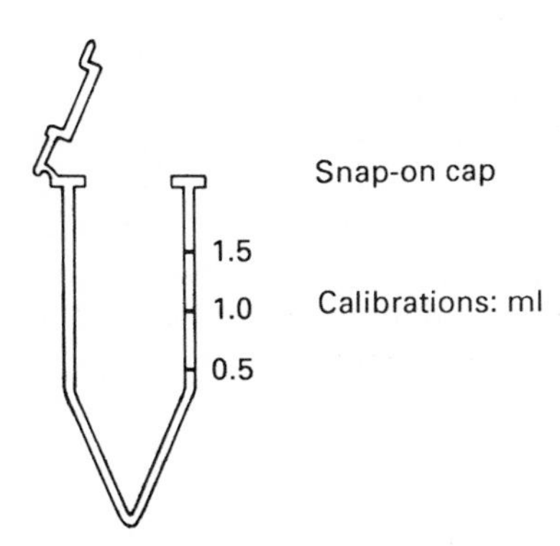

Fig. 9 An Eppendorf plastic tube, commonly used for processing and storage of small volumes of reagents and specimens, in conjunction with a microcentrifuge. The tubes are also available with a more reliable screw cap at considerably increased cost.

subcellular fractionation or molecular purification. It may be of intermediate performance, developing *g* forces up to perhaps 100 000, or high performance and able to operate at several hundred thousand *g*. Such an instrument can cost up to £50 000 with a full range of rotors. These larger machines will often require a three-phase power supply and cooling water. Rotors revolve in a vacuum to cut down heat generation by friction with the air. Finally, miniaturization of ultracentrifuges is also a useful current trend. Benchtop machines delivering 100 000 *g* and powered by compressed air can be used for volumes up to 200 μl, and can pellet subcellular particles in minutes rather than hours. For volumes up to a few millilitres per tube, a benchtop ultracentrifuge is now available which also has some of the same advantages of small scale as the microfuge.

SPECTROPHOTOMETERS

These are versatile, and in anything more than a basic teaching laboratory, essential items. They range from a basic colorimeter to computerized scanning and recording instruments.

Colorimeters are based on the use of filters to provide incident light of different wavelengths. Filters tend to transmit a fairly broad band of light, so are not particularly well defined or reproducible, but colorimeters are adequate for basic measurement of bacterial growth in liquid (i.e. turbidity), and for elementary enzymology using chromogenic substrates. Often tubes rather than optical cells are used to hold samples.

For greater reproducibility and accuracy, a *spectrophotometer* is needed. Preferably it should hold either tubes or cells of 1-cm pathlength, as well as mini-cells for small volumes. Again, electronic circuitry and pushbutton operation are greatly simplifying and speeding up use of these instruments. Good basic instruments, manually operated for single samples, can be bought for several hundred pounds. For greater versatility, ultraviolet absorbance capability can be added for at least 50 per cent increase in price. Protein and nucleic acid levels as well as a wide variety of enzyme reactions linked to NAD/NADH (absorbing at 340 nm) can then be monitored directly.

One development which is worth a mention is the *microplate* or *ELISA reader*. This is essentially a multichannel colorimeter, designed to read simultaneously blocks of eight wells in a 96-well microtitre plate. The entire plate can be read, and a print-out on thermal paper strip of absorbance values for all 96 wells produced in about one minute. The system exploits optical fibre technology to deliver incident light from a single source to each of eight wells simultaneously, and a block of light-sensitive cells to monitor transmitted light. The plate is moved automatically to scan all wells in blocks of eight. Although designed originally for use in enzyme-linked immunoassays, these instruments are very versatile especially for miniaturized tests in, for example, enzymology, and in a class-teaching situation where large numbers of readings often have to be performed simultaneously. The equipment is expensive though – several thousands pounds for a good quality instrument.

OTHER MISCELLANEOUS EQUIPMENT AND FACILITIES

A *vacuum pump* and/or *compressor* may be useful, particularly for filtration and (vacuum) drying processes. Gases to be evacuated must be non-corrosive – volatile organic acids may quickly corrode pumps or contaminate the pump oil. A good vacuum pump is essential for operation of a freeze drier – also essential in a well-found laboratory both for the preservation of cultures and for concentrating labile solutes from aqueous solution.

Glassware of various kinds can be considered as equipment, although as disposables are introduced increasingly, we can regard these items more as consumables (and have to budget for them!). The basic glassware – beakers, flasks and measuring cylinders in a variety of sizes – is common to all biological laboratories. A good range of conical flasks is particularly necessary in microbiology for use as shaken culture vessels. There is also a need for large numbers of screw-cap bottles and tubes, which are still routinely made of glass and recycled, although almost all are now available (at a cost!) as plastic disposables. Bottles are invaluable for medium sterilization and storage, and as culture vessels. Useful sizes are the traditional ¼ oz or 'bijou' bottle holding about 5 ml, the 'universal' 1 oz bottle which holds about 25 ml, the 'medical flat' 4 oz bottle (100 ml), and 500-ml bottles for storage of solutions. Test tubes, particularly 125 mm size, are also useful and if fitted with loose aluminium caps they can be autoclaved easily and stored sterile.

Storage racks for bottles and tubes are essential, and often in short supply in a busy lab – again we are up against a 'Parkinson's law' which means that the specimens expand to fill all available racks! Racks are needed especially for incubation or cold storage of cultures or extracts, and are preferably made of metal with a corrosion-resistant coating. Baskets, similarly made, for storage and carriage of, for example sterile bottle stocks, are also essential. For plastic tubes such as the Eppendorf type, plastic racks are available, and insulated boxes of expanded polystyrene can also be obtained – they are ideal for freezer storage and will conserve low temperature when tubes are removed for sampling, for example of cultures or enzyme solutions kept liquid with glycerol. For low budget work, such racks and holders can be improvised easily from expanded polystyrene packaging with holes pushed through – they will even provide convenient floats for incubating Eppendorf tubes in water baths (Fig. 10).

Graduated pipettes are also becoming less used as pipetting devices with disposable tips become more versatile, but there is still a need for them in most laboratories. They should be sterilized in metal containers with accurately fitting caps which stay on by friction, but do not jam! Useful pipette sizes are 1, 2, 5 and 10 ml, as well as pasteur pipettes holding about 2 ml. Conventionally, pipettes are plugged with a fairly loose

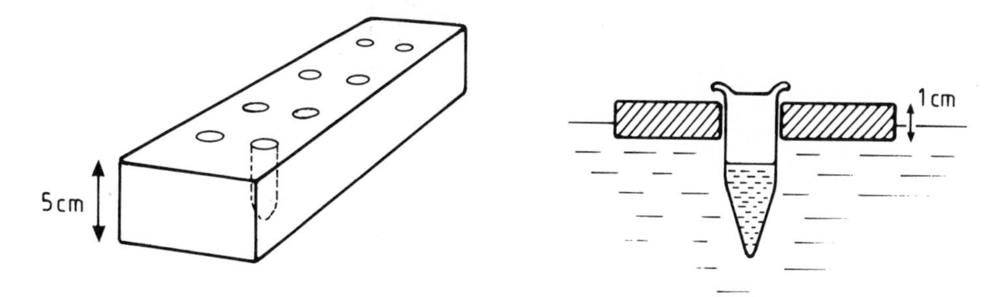

Fig. 10 A rack and float for Eppendorf tubes, easily made at no cost from expanded polystyrene packaging material, and used for storage or handling and for water bath incubations, respectively.

cotton-wool plug (so that it can be removed reasonably easily for cleaning), and sterilized in an oven.

A range of reliable and easily operated pipetting devices should be available to deter users from mouth pipetting. Simple ones are usually more reliable, although the traditional rubber bulb with ball valves is too easily damaged by accidental aspiration of liquid into the valves. Models based on a plunger system like a syringe are often durable and reliable.

Micropipettes are now indispensable in any other than basic teaching environments. A wide range is available, but for reliability, accuracy and durability, it is unwise to buy the cheapest. Adjustable models are more versatile than fixed volume although the latter may be adequate for routine identical tasks performed every day. They should accept standard tip sizes or shapes, and if potentially hazardous organisms or solutions are handled, a tip ejector mechanism is a great advantage. Useful volume ranges are 2–20, 20–200 and 200–1000 μl. Good quality micropipettes cost about £100 each.

Filtration is an important process in microbiology, especially for sterilization of media and solutions. Although pre-packed and pre-sterilized, disposable filtration units are now commonly used for small-scale work (see below), self-assembly units with replaceable membranes remain important for larger volumes. Asbestos pad 'depth' filters (Fig. 11) were formerly used extensively for sterilization of complex media (see Chapter 5), but are now known often to adsorb and remove from media appreciable amounts of potentially important growth factors. Depth filter pads were usually used with a sterile side-arm flask and vacuum pump to provide a pressure difference across the filter, or for larger volumes a pressure vessel and compressor (Fig. 12). Similar methods can be used with membrane filter units (see Chapter 5), again often in self-assembly units which can be sterilized, for example by autoclaving. For small volumes, syringes can be used to generate filtration pressure, but for large volumes pumps are required.

Various means of physically breaking up microbial cells are commonly employed at the research level. High-speed *homogenization* generates surface shearing forces capable, for example, of breaking off bacterial flagella. An ordinary domestic blender may be adequate for some uses, and simple high-speed laboratory blenders are

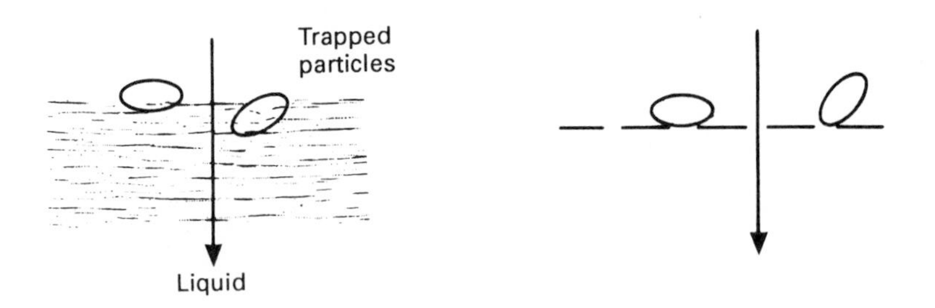

Fig. 11 Types of sterilizing filter. The depth filter, built up from multiple layers of fibrous material, relies on the trapping of particles at, or below, the surface of the filter, with increasing probability of entrapment as the liquid penetrates the filter. The more recently developed membrane filter is a single layer of material, with no holes large enough to allow any penetration of the particles being filtered.

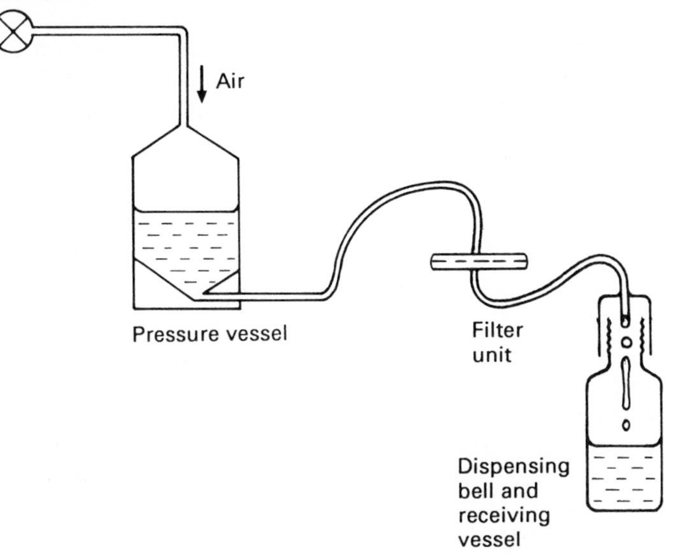

Fig. 12 A pressure filtration system for large-scale use. A compressor, pressure vessel and dispensing mechanism for aseptic delivery of the filtrate are required. On a smaller scale, disposable plastic syringes are routinely used to force liquid through small filtration units.

relatively cheap. More versatile models, for example with caged rotors and sealed homogenization vessels to prevent escape of aerosols of infectious material, are more expensive and may cost several thousand pounds for the most sophisticated.

Powerful shearing forces as well as sudden pressure changes are also generated by presses, which can break open microbial cells, and are based on extrusion of a concentrated mass of cells through a small orifice under extremely high pressure.

Possibly more versatile is *ultrasonication*, now widely used for both removal of surface components and rupture of cells. In order to generate the necessary shearing forces, a vibrating probe must be immersed in the cell suspension – bath sonication of the container is not usually sufficient. Larger probes and more powerful machines are needed for large volumes. A general-purpose probe sonicator for small-scale laboratory use will cost several hundred pounds.

For gentler agitation of cell suspensions or re-suspension of pelleted cells, a *vortex mixer* is essential. A good quality durable model with adequate power, variable speed and automatic pressure switch will cost around £100 – a good investment in a busy laboratory. Magnetic stirrers are also essential for gentle agitation, particularly in making up chemical solutions to save the operator stirring time! Again, these are less than £100, or around that figure with a built-in hotplate.

A general-purpose oven, preferably with fan, is very useful for drying glass or plastic ware. An ordinary domestic microwave oven is also a boon for boiling small quantities of liquid, especially agar solutions which are easily burnt if heated by bunsen burner, for example.

Many other small miscellaneous items will be needed, and their cost will not be negligible. *Washbottles* are useful not only for distilled water but also for disinfectant ready to use, and for saline, solvents (methanol for fixing cell smears for microscopy, acetone for cleaning glassware), buffers and other solutions constantly in use. Washbottles should be of good design and robust construction – cheap ones can be very frustrating! *Thermometers* (glass) are frequently needed and frequently broken, and again electronic probe devices are available and should be considered as a more rugged alternative. Gas lighters can be a constant source of trouble – piezo-electric devices operated by pushbutton are probably best. As in every laboratory, tripod, gauze and retort stands will be needed occasionally.

ANAEROBIC FACILITIES

If the laboratory will be involved in work on anaerobic or microaerophilic organisms, facilities for incubation in low oxygen environments will be needed, and for very strict anaerobes these may need to be quite sophisticated. Oxygen-tolerant organisms, e.g. many *Clostridia*, and microaerophiles, can be handled adequately in air and simply incubated in anaerobic jars. Formerly these were commonly of the Mackintosh and Fildes type, with two valves for replacement of the atmosphere with hydrogen and a catalyst to ensure reaction of residual oxygen with the hydrogen (Fig. 13). For safety and convenience, these have been largely replaced by chemical systems such as the GasPak (Becton Dickinson), which can be used in simple sealed containers holding about 12 standard petri dishes. Special packs are available to generate a microaerophilic atmosphere enriched with CO_2. The latter atmosphere can also be generated very simply and cheaply in a candle extinction jar – a sealable metal container in which a lighted candle is allowed to burn to extinction.

For work with organisms intolerant of oxygen, more complex equipment is needed. Much can be done with specialized culture vessels, bottles or tubes with butyl rubber bungs or screw caps, which are purged of air during all manipulations by a stream of oxygen-free, sterile gas. All media are prepared oxygen-free and kept pre-reduced. This

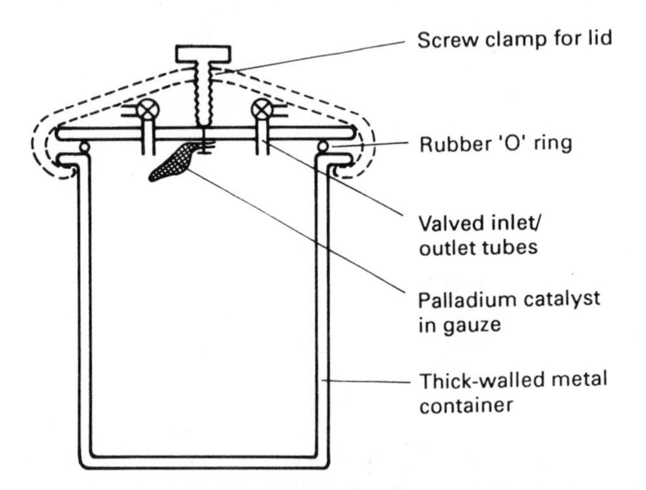

Fig. 13 An anaerobic jar of the Mackintosh and Fildes type, with inlet and outlet valves for evacuation and input of hydrogen, and catalyst for purging of residual oxygen. The valves are unnecessary with the gas-generating pack now commonly used, and the jars are often made from transparent plastic.

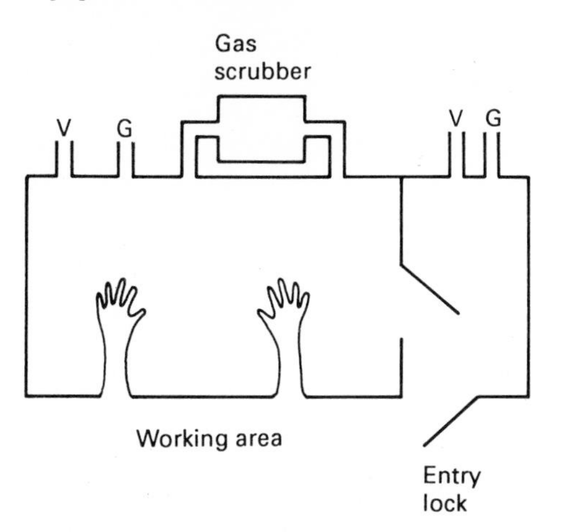

Fig. 14 Diagrammatic representation of an anaerobic cabinet. The construction may be either from flexible plastic film, or as a rigid box with transparent windows. Normally an oxygen-free gas mixture with a small proportion of hydrogen for catalytic oxygen purging in the scrubber, to which the atmosphere is continually recirculated, is supplied from a cylinder at G. The vacuum lines V permit either evacuation of the entry lock, or removal or flushing away of air during initial filling of the chamber.

is the 'Hungate' technique, and equipment needs are fairly simple – mainly suitable gas supplies and columns for catalysts to purge traces of oxygen, tubing, valves, filters and manifolds for gas delivery, and a simple electrically driven roller system for rotation of tubes and bottles.

Alternatively, a complex and expensive *anaerobic glove box* can be used. This is essentially an entire workstation contained within an artificial atmosphere. The basic requirements are shown in Fig. 14. Its use will be described in Chapter 8. Such an item may cost several thousands pounds, and is also expensive in use due to the large requirement for compressed gasses.

Materials

There is a continuum between essentially disposable materials and re-usable equipment, and increasingly, the former items are displacing the latter. In this section I will try to cover both.

BASIC MICROBE HANDLING

The most traditional and essential tool for the microbiologist is the *bacteriological loop*. This can be made of wire, usually of an alloy of nickle and chrome (nichrome) for cheapness, but ideally of platinum, and mounted in a holder. Straight wires or needles are also useful for precise sampling of individual colonies. Alternatively, disposable items such as toothpicks or purpose-made pre-sterilized plastic loops can be used. For examples, see Fig. 15.

For sampling wet environments, particularly in medical work, cotton or alginate

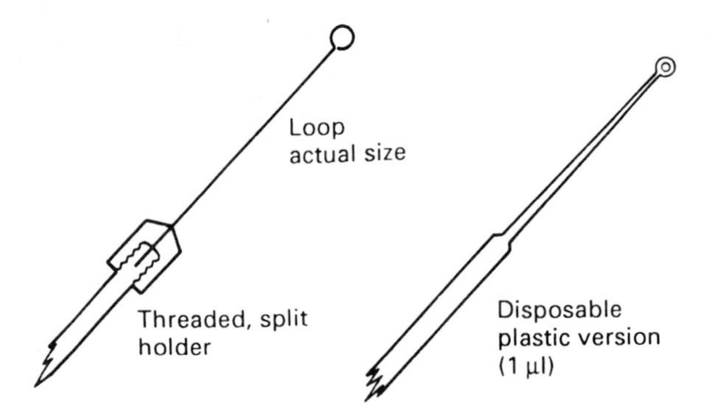

Fig. 15 Loops for handling of cultures. The wire loop should be accurately made to about the size and shape shown. Plastic disposable loops are more expensive to use but may be convenient, for example in anaerobic work inside a cabinet.

swabs are useful. They can be 'home-made' by winding a small amount of absorbent cotton-wool smoothly round a small wooden 'orange' stick and autoclaving in test tubes, or bought ready sterilized and pre-packaged.

Microscope slides are also a significant consumable item. They are quite cheap, and can be re-used or discarded. If re-used they are quite laborious and difficult to wash properly, and will soon become scratched, so this may not be cost-effective. Ordinary slides for disposal need not have ground edges. Their thickness may be important for some types of microscopy. In addition cover slips will be needed for wet mounts or brightfield examination with a $\times 40$ objective (an oil immersion lens will not require a coverslip). The most convenient size is 18 mm square.

For staining smears, some kind of slide support is needed. For small numbers of slides, a 'bridge' of two parallel glass rods across a sink (Fig. 16) is traditional and effective. Stains are kept in dropping bottles. For large-scale work, slide racks and staining baths can be used.

One item which should not be disposable (but is often broken!) is the *counting chamber* for microscopic counting of particles. One type, the *haemocytometer*, is

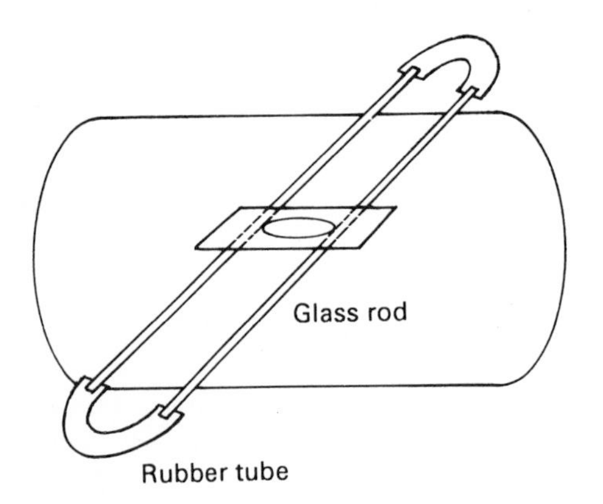

Fig. 16 A simple staining bridge for microscope slides, which can be arranged over a sink or basin to receive waste or spills of stain and rinses.

suitable for animal cells such as blood cells and for yeasts, but not for bacteria since the depth of the chamber (0.1 mm, or 100 μm) is so much greater than the dimensions of a bacterial cell that many minutes would be required for all cells to settle into the same plane for high-power microscopy. Instead a 'Thoma' or 'Helber' chamber is used, with a depth of 0.02 mm (20 μm), and the slide is thin enough for phase-contrast microscopy (hence it is very delicate, and being accurately ruled at 50-μm intervals as well as accurately ground for chamber depth, rather expensive). Use of these chambers is described in Chapter 9.

CULTURE VESSELS

The most basic vessel is the *petri dish*. Formerly made of glass, disposable plastic is now used almost exclusively, at a cost of a few pence each. There are many suppliers, who compete keenly and keep prices down. Bulk buying will often lead to discounts, and if storage is available the dishes will keep indefinitely. To prevent a moisture seal forming between base and lid, small projections are moulded into the plastic at the angle of the lid (Fig. 17) to create a small ventilation gap. There may be one or three (single or triple vents) according to the need for air movement. If problems occur through accumulation of moisture during incubation, triple-vented dishes may be useful.

Other plastic containers such as tubes or screw-cap bottles may also be useful for culturing, although glass is often used. Disposable tubes of various kinds for handling organisms are also available.

CHEMICALS AND MEDIA

Standard laboratory *chemicals* of good quality are widely available. For routine use, for example in preparation of undefined media, the highest purity chemicals need not

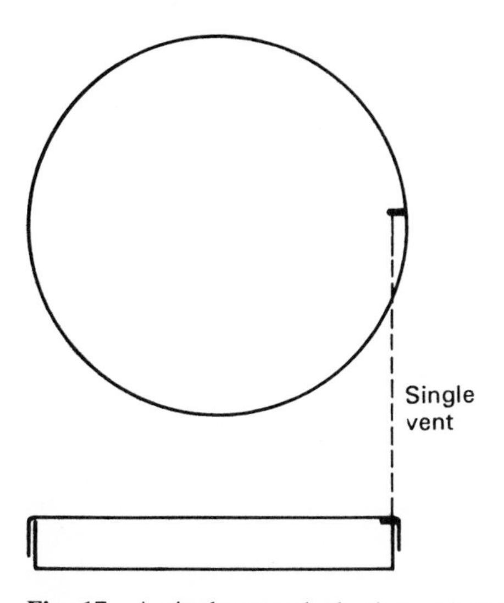

Fig. 17 A single-vented plastic petri dish. The vent is a small projection of the lid which prevents its complete contact with the base. Dishes may have either single or triple vents, depending on the required extent of gaseous exchange with the environment.

Table 2 Basic facilities and equipment needs for different types of laboratory. Lists are not exhaustive but will give some guidance at a glance on the scale of equipment needs. Items for the more complex lab are additional to the basic needs.

Facility	*Elementary teaching*	*Routine/small scale*	*Research/large scale*
Sterilization	Pressure cooker	Benchtop autoclave. Sterilizing oven	Floor-standing autoclave
Discard		Autoclave or incineration	Organized system
Microscopy	Elementary, to × 400, brightfield	Phase contrast, to × 1000; plate microscope	Fluorescence. Inverted
Incubation and heating	Incubator. Bunsen burner or portable burners	Variety of incubators. Water baths. Piped gas	Warm room. Heating blocks
Cold storage		Fridge, freezer	Cold rooms, lower-temperature freezer
Workstations	Open bench	Clean room if practicable	Sterile air bench, safety cabinet
Balances	Rough balance	Fine balance	
pH Measurement	Indicator paper	Hand-held electronic	Research instrument
Centrifuges		Benchtop – small and micro	Full range including large volume free-standing refrigerated and ultracentrifuge
Spectrophotometry	Colorimeter	Simple and rugged variable wavelength (visible light)	Scanning-recording and UV; microtitre plate reader
Gas handling		Vacuum pump/line. Compressor	
Miscellaneous		Vortex mixer. Drying oven. Magnetic stirrer. Desiccator. Micropipettes	Sonicator. Homogenizer. Microwave oven

be used, but in many cases the cost of analytical-grade chemicals is not much greater and it is convenient to keep stocks of one grade only. Chemicals (other than dangerously toxic or corrosive items – see Chapter 5) should be stored where they are easily accessible for regular use – preferably near balances – and in, for example, alphabetical order for easy location.

Media are also widely available from commercial sources. Most commonly used are dried, powdered media which can be dissolved and sterilized in the laboratory. Such preparations are convenient to transport and store, and relatively cheap to use. They do require however that facilities are available for sterilization, and in the most elementary teaching laboratory it may be necessary to buy pre-prepared, sterile liquid media and pre-poured agar plates, which can also be obtained. At the other extreme, the largest laboratories may prepare some of their own media, such as nutrient broth, in a 'media kitchen', but variability in the product and the expense and inefficiency of the facility may be real problems and this practice is becoming rare.

WATER

A supply of distilled water is essential. For routine preparation of media and simple buffers etc., an ordinary glass still provides adequate purity. For more exacting work, for example enzymology or molecular biology, double-distilled water will be needed and even this may not be good enough for, for example, tissue culture work. Additional filtration and ion exchange may be needed, followed by autoclaving if the water is to be stored or used in DNA manipulation. The economics of pure water production may in fact dictate that ion exchange is used in preference to distillation, due to the high energy cost of the latter.

Equipping a microbiology laboratory

If facilities for microbiological work are being set up from scratch, it may be useful to see at a glance the level of equipment necessary, with major items of expenditure, for a variety of levels of work. I have therefore summarized the major items discussed above in Table 2.

CHAPTER 3

Microscopy

Microscopy is central to microbiology, but it is often not used very effectively. To be done well, it requires technical skill and knowledge of the instrument. There is nevertheless a tendency for expert microscopists to regard the tool as an end in itself, giving it a mystique which can be a barrier to the novice user. The results of misuse are often depressingly uninformative, and there is also an element of learning by the eye or brain, to interpret images correctly, which only comes from successful practice. Much can be done, however, by following a few simple rules, to enhance the value of simple microscopy.

The specimen

Broadly speaking, there are two types of specimen the microbiologist will examine frequently: (1) dried, fixed and stained smears, and (2) wet mounts of unstained material. Basic rules are summarized below for ready reference.

- Stained smears are examined by brightfield.
- Mounting in, for example, oil, with a coverslip, is necessary with a $\times 40$ objective.
- Oil immersion between the specimen and objective is necessary with a $\times 100$ objective. The coverslip may then be omitted.
- Oil immersion must not be used with a non-immersion objective – in general objectives less than $\times 100$.
- Wet mounts are examined by phase contrast (or darkground – see below); a coverslip is always required.

In general, specimens examined in microbiology will be either stained smears or wet mounts. Stained smears will require mounting under a coverslip if they are to be examined with a $\times 40$ objective, and this can be done in a drop of immersion oil if they are to be discarded. Alternatively, for examination with a $\times 100$ objective, oil can be placed directly on the specimen which is examined without a coverslip. For wet mounts, it is important that the right amount of liquid is introduced under the coverslip – if too little, it will not completely fill the space and capillary action will cause it to flow across the field of view as it spreads, so that a static view of the specimen is difficult to obtain; if too much, the coverslip will float on the liquid and move about with the slightest vibration of the bench. Examination of wet mounts with a × 100 objective and oil immersion is not easy – surface tension tends to lift the coverslip with the immersion oil as the microscope is racked up. Certainly for convenience it will be preferable not to use an oil-immersion objective.

The preparation of stained specimens is thoroughly covered by numerous other authors, especially in the field of clinical bacteriology where identification of organisms depends in part on their staining properties. For routine use, the simple Gram stain is most useful – normally performed on smears heated gently to fix them, by passing several times through a bunsen flame until just too hot to touch for more than an instant on the back of the hand. For microscopy, smears should ideally not be too thin or too thick, and if difficult to find and focus on, the edge of the slide or coverslip or a visibly heavily stained area may be a good starting point. Methods for Gram staining are variable, but all depend on staining with crystal or methyl violet (1 min), followed by Gram's iodine (1 min), decolorization with ethanol, and counterstaining with a pink stain such as saffranine or Rose Bengal (1 min), safer than the carbol fuchsin originally used and now believed to be potentially carcinogenic. The film is best blotted gently or drained dry after rinsing. The exact timings of each stage of the procedure may need to be varied to optimize for particular stain preparations. For 'instant' staining if determination of Gram character is not critical, staining with crystal violet for a minute or two followed by a water rinse and drying is adequate. Many other specialized staining methods are available, for example to demonstrate the presence of bacterial endospores.

For the best results, both slide and coverslip should be new. Smears such as fingerprints can be removed before use by breathing on the glass and wiping vigorously with a clean paper tissue, and then blowing away dust before use.

Focussing

This is probably the trickiest procedure at first, since at high magnifications the depth of field is small and if the stage is racked up and down rapidly, the image may come and go rapidly and not be noticed, especially if organisms are sparse. Fortunately, however, the objectives are usually all designed to be in focus at the same distance from the object, so if the microscope is focussed using a low power objective on (e.g. the edge of the coverslip), the objective can then be changed to a higher power lens which should be close to focus.

Initially it may be useful to focus on an easily seen object, such as a grease pencil mark, on the brightfield setting of the condenser. The eyepieces can then be adjusted separately so that both are correct for the user. If only one is adjustable, focus the microscope as well as possible for the fixed eyepiece, and then adjust the other for the best performance with the other eye.

The microscope

For the beginner, the basic controls are listed below. The layout of the instrument parts is shown very diagrammatically in Fig. 18.

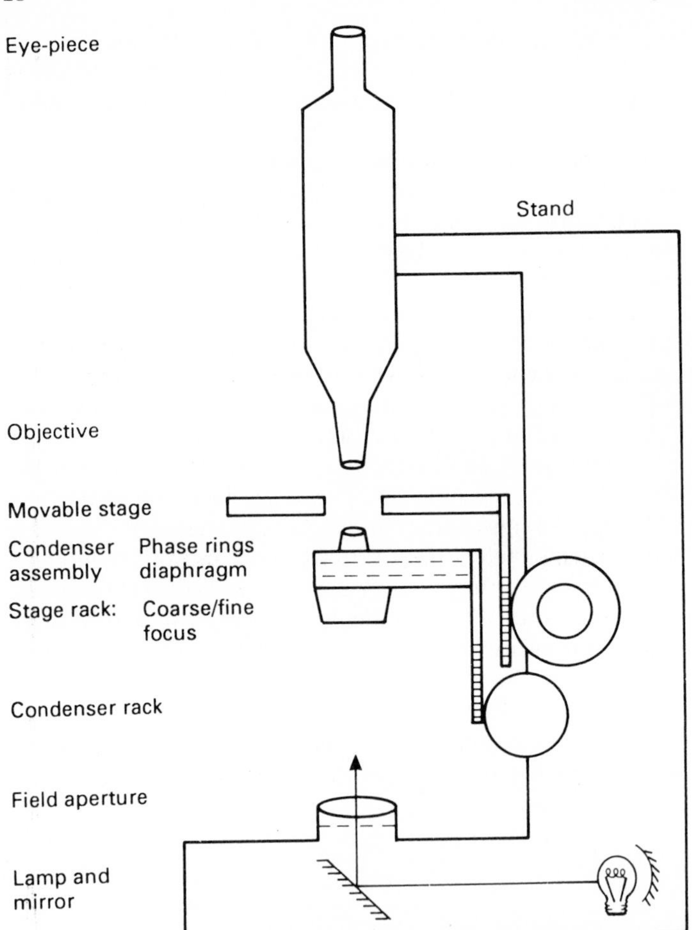

Fig. 18 A typical modern microscope, illustrating the main parts referred to in the text.

Focussing Normally the stage is racked up and down on a modern instrument, usually with coarse (outer) and fine (inner) concentric knobs on either side. Do not confuse with the condenser rack.

Condenser selection A revolving disc or sliding cartridge brings different lens or phase ring systems into the light path beneath the specimen, depending on the need for brightfield, phase contrast or darkground.

Condenser position A separate rack system is used to raise and lower the entire condenser assembly beneath the stage for optimum performance.

Condenser diaphragm A lever or revolving ring is used to alter the aperture of the condenser in its brightfield position, leading to subtle variations in the contrast of the image. The adjustment of this iris is also critical for setting up Kohler illumination (see below).

Some microscopes have a device for temporary removal of the upper element of the condenser lens, usually by swinging it sideways out of the light path, so that a larger area can be illuminated for very low power work on brightfield.

Field diaphragm A diaphragm in the base of the microscope, which is normally wide open for routine use. It can be used during focussing of the condenser for Kohler illumination.

Objectives These are normally selected by rotation of a turret holding several objectives. In general, the main requirement for their good performance is a perfectly clean surface.

Eyepieces These are adjustable for distance apart, and one or both of them also for focus depending on the eyes of the user. Some instruments may have provision for adjustment of eyepiece focussing depending on inter-pupillary distance which can be read off a scale (around 60–70 mm) after adjustment to suit the individual. The eyepieces can then be adjusted to a similar point on the scale around their circumference – the reason being that sideways movement of the eyepieces slightly alters the tube length, requiring slight alteration of focus.

Microscopy

There are two main aspects to routine microscopy of microorganisms: (1) examination of stained smears, and (2) examination of wet mounts.

Examination of stained smears This will usually be done either to identify uncharacterized cultures, or to verify the identity of a culture about which there may be doubt, for example whether it is contaminated. In both cases as much detail as possible of the morphology of the culture will need to be seen, and it is most effective to go immediately to the × 100 oil-immersion objective, on brightfield. The condenser will have to be at its highest position beneath the specimen, and the condenser diaphragm initially wide open. If a series of cultures is being examined on identical slides, the rack position at focus will be approximately the same for each. To change slides, swing the objective turret to an intermediate position so that the objectives are well clear of the slide, and the focus need not be moved. If focus is lost, bring the objective as close as possible to the specimen without actually touching it (inspect from the side), then rack slowly away from the specimen while watching for the image to come and go. Do not try to focus *towards* the specimen – even though there is normally a spring mount in the objective lens to prevent damage if it is racked into the slide, it is surprising how quickly this may reach the backstop! If you cannot find focus, move to a denser area or a felt-pen mark at the edge of the smear. Having located the specimen, check around several areas of the slide that the staining is even and identical all over, and you have not found an atypical area.

Examination of wet mounts For this purpose, phase contrast or dark ground will be needed. To avoid the necessity for oil between the condenser and slide, phase contrast may be preferable and, if of good quality, is almost equally informative. Ideally, for phase contrast the microscope should be set up for Kohler illumination, in which the condenser is positioned so that the image of the field aperture in the base of the microscope is in the same plane as the specimen, whereas the image of the light source is not in that plane but rather in the plane of the condenser aperture. This is achieved by focussing on a specimen, closing the field aperture almost completely, adjusting the height of the condenser assembly until the image of the field aperture appears in focus coincidentally with the specimen, and then opening the field iris again. This use of the field iris is also ideal for centring the condenser – so that the image is central to the field of view. Having made the necessary adjustments to the condenser height and field iris, remove the eyepiece and look down the tube. A small, focussable telescope may be provided for this. The image of the condenser iris should now be visible, and this should now be adjusted to fill about 70 per cent of the field of view. Next, the appropriate phase contrast position of the condenser should be selected, depending on the objective to be used. On some microscopes, phase contrast requires adjustment of phase rings in the condenser to align them with the objective. This is done with two adjustments working at right angles, so that the images of the rings in the condenser and objective, seen by removing the eyepiece and looking down the tube, coincide (Fig. 19). Again, the telescope may be needed to focus on the images of the phase rings during this operation. Replace the eyepiece. Provided that the objectives and the upper surface of the condenser are scrupulously clean, good contrast should then be obtained with an unstained, wet-mounted specimen.

On cheaper microscopes, and on some dearer models, many of the above adjustments are pre-set and routine use of phase contrast is simpler. The technique is very useful for rapid monitoring of cultures to check morphology or viability (by motility), for contamination of cultures or even tissue cultures, or for example to monitor cell breakage perhaps after sonication. The lack of need for staining greatly increases its convenience, and it is worth mastering the intricacies of the microscope so that it can be used routinely.

There are several common mistakes which beginners make that lead to very puzzling results. In brightfield work, the substage condenser may be closed down too far to allow sufficient light to pass. The condenser top lens may be left out when using higher power objectives. Brightfield may be used with wet mounts, or phase contrast with stained preparations. Bacteria may be looked for at too low a magnification for them to be visible. All these errors may lead, frustratingly, to formation of a very strange image or no image, and the correct conditions are therefore tabulated in Table 3.

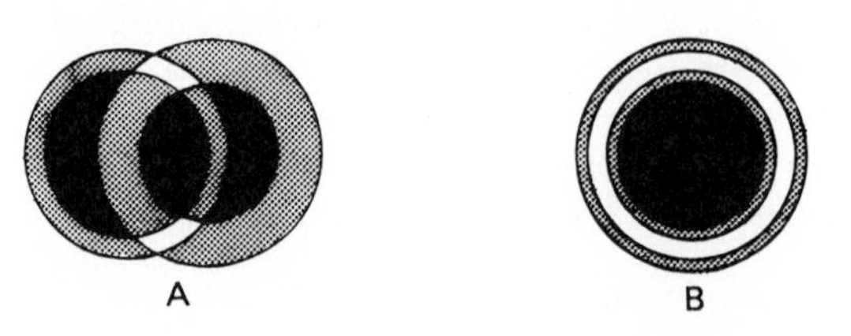

Fig. 19 Adjustment of phase ring in the condenser assembly so that the rings in the condenser and in the objective, out of alignment at A, are concentric as shown at B.

Table 3 Summary of conditions for simple microscopy.

Mount	*Stain*	*Phase contrast*	*Objective*			
			× 10	*× 25*	*× 40*	*× 100*
Dry	+	−	Swing-out condenser top lens	Bacteria barely visible	Coverslip needed	Oil
Wet (coverslip)	−	+				Oil

DARK FIELD MICROSCOPY

Dark field microscopy at low magnifications may often be possible by using the 'wrong' phase contrast objective/condenser phase ring combination. For best results, however, a purpose-made dark field condenser is necessary, and for high-power work, a special objective with a diaphragm in the back focal plane will be necessary. The condenser can be centred by adjusting its height on low power until a dark circle appears in the field of view, centring this, and then moving to a higher power objective and adjusting the condenser height for maximum brightness and contrast. The main problem with dark field is that bubbles, either in the oil below the slide, in the mountant or in the immersion oil above the specimen if high power is used, can produce glare even if they are well away from the specimen. Therefore, scrupulous attention to the preparation of the slide is necessary, and frequent renewal of the oil will help to avoid build up of bubbles. If they do occur, they will float, so wiping either the underside of the slide or the objective after racking up will usually get rid of them.

FLUORESCENCE MICROSCOPY

Fluorescence microscopy in microbiology is usually the final stage in a complex series of manipulations to stain specifically with antibodies, or other probes, specific molecular structures of the microorganism (see Chapter 12). The microscopy itself is fairly straightforward. Ideally, the specimen is first focussed and examined by an alternative system, most usefully phase contrast, so that major features can be identified. Fluorescence conditions are then adopted, and commonly either trans-illumination or epi-illumination may be used, traditionally with a mercury vapour light source to provide ultraviolet illumination. These lamps are however potentially dangerous, since they may explode if switched on and off without attention to warming and cooling periods, or if run for longer than their recommended lifetime. It is therefore preferable, for use with fluroscein at any rate, to use tungsten–halogen lamps which excite fluorescein well in the blue area of the spectrum, although possibly with marginal loss of brightness of the fluorescence. It will be found that fading of fluorescence occurs quite quickly unless agents such as DABCO (1,4-diazo-bicyclo-2,2,2-octane) are used to help prevent it. Recording of results by photography may in any case need very long exposures.

ELECTRON MICROSCOPY

While it is not the function of this book to provide any detailed information on electron microscopy, in case there is access to an electron microscope it is worth emphasizing the usefulness of just one method in microbiology. This is *negative staining*, which is a very simple, rapid and convenient technique requiring no training other than in the microscopy itself! It also tends to be much under-exploited.

The method relies on the deposition of electron-dense stain from solution, around objects supported on an electron-transparent film, usually of formvar. The method relies on high solubility of the stains used, so that at concentrations which are adequately electron dense to provide contrast in films only nanometers thick, the stain essentially does not crystallize during drying: if it did, spurious detail would be introduced into the image due to the uneven deposition of stain. For the same reason, other solutes liable to crystallize on drying, i.e. salts, must be substantially excluded from the preparation. Stains which are often used include phosphotungstate and molybdate, with high atomic weight metals for maximum electron density.

For negative staining, bacteria from, for example 1 ml of overnight culture, are pelleted in an Eppendorf tube, the supernatant removed, and the pellet (or a small amount of agar growth on a loop) resuspended in two drops of distilled water. One drop of 2 per cent potassium phosphotungstate (pH 7.2) is immediately mixed in, and a drop transferred to the upper (coated) surface of a formvar-coated specimen grid picked up with watchmaker's forceps (Fig. 20). Excess liquid is drained rapidly by touching to a filter paper, and the film dried rapidly by waving it in the air. Speed is necessary at all stages to prevent osmotic disruption of the cell structure, since salts and other solutes are absent. It is then ready for examination. Bacteria should be seen individually distributed – if not, another grid can be prepared with a thinner suspension. Objects surrounded as they dry by a film of stain will be picked out as paler images against a dark ground – the larger the object, the darker the background (Fig. 20). It is a particularly powerful but simple technique for observation of, for example, bacterial flagella or fimbriae.

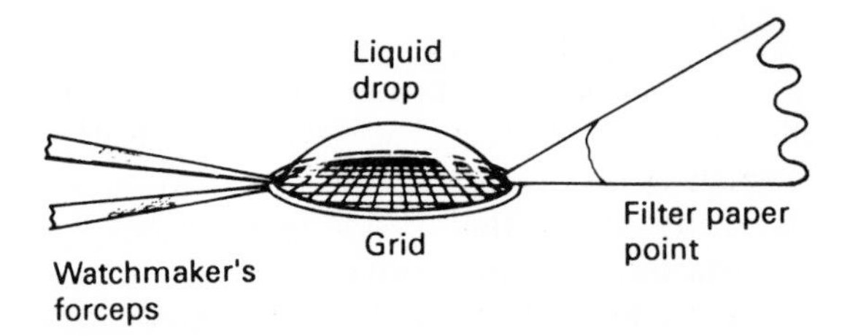

Fig. 20 Application of a specimen in suspension onto an electron microscope grid. The grid is held at one edge with watchmaker's forceps, a drop of suspension applied with a pasteur pipette, and surplus liquid drawn off by touching to a filter paper point.

CHAPTER 4

Media and growth

Types of media

The main categories of media are most usefully described as *complex* and *defined*. The former is a complex mixture of nutrients, salts and other solutes obtained from a natural source such as meat or milk, and therefore essentially slightly variable in composition however carefully controlled the methods of preparation. The latter is a mixture of known, purified constituents in known proportions and concentrations, which in theory should be reproducible from batch to batch and from lab to lab. Complex media are cheaper than defined media, and often support better growth, especially of fastidious organisms, i.e. organisms which have complex and sometimes incompletely known growth requirements. Lack of reproducibility may however limit their usefulness in research.

Other descriptions of media which may be encountered are simple, minimal, synthetic and semi-synthetic. *Simple media* imply a single main constituent such as peptone, a mixture of amino acids obtained by hydrolysis of protein, without additional supplements necessary for fastidious organisms. *Minimal media* are basic mixtures of essential salts with carbon and nitrogen and energy sources, and may support growth of organisms which have the ability to synthesize all their own organic constituents (although the energy cost may be high, leading to lower growth rates and yields than those obtainable on complex media). *Synthetic media* are essentially the same as defined media, but the word is a little less precise, and reproducibility may not be total; for example they may include an amino acid mixture free of protein or large peptides, obtained from protein hydrolysates which have been passed through a dialysis membrane. *Semi-synthetic media* are hard to define as well as of limited scientific value: usually they consist of a defined or synthetic base with the addition of a complex supplement such as yeast extract, which negates the value of the defined base.

Basic components

NUTRIENT BROTH AND NUTRIENT AGAR

These are the traditional, cheap, complex media for non-fastidious organisms such as *E. coli*. For routine isolation and identification procedures they are adequate, but they have been superseded for molecular biological studies, for example, by hydrolysate-based media.

AGAR

Agar is the routine gelling agent for solid media, and has many advantageous properties. It has unusual melting and setting properties which are usually useful but are puzzling to the beginner: it melts only at 100°C, but sets at much lower temperatures, around 45–50°C, although lower setting point agars are available. This means the gel will not melt even at elevated incubation temperatures (gelatin, for example, is normally liquid at 37°C or slightly above), but it also means that if the agar cools a little too much before plates are poured and starts to set, it cannot be re-melted rapidly by applying a little heat. Agars vary in their melting points, gelling strength, purity and acidity. Semi-purified preparations such as ionagar may be recommended for certain specialized media, for example to provide a high gel strength with minimal agar concentration. The refined carbohydrate component of agar is agarose – highly expensive compared with agar sold for media, but very uniform in composition and useful for critical work with minimal media. It is widely available as an electrophoresis support for macromolecular analysis.

For some purposes, a softer-than-normal agar gel is required. Normally for gels strong enough to withstand streaking with a wire loop, concentrations around 1–1.2 per cent are used. As little as 0.1 per cent will however set to a semi-solid state sufficient to prevent convection, which may be useful in some circumstances.

BASIC COMPLEX MEDIA

Various highly nutritious complex extracts of animal or plant tissues are used as bases for complex media. Examples are brain–heart infusion broth and malt or potato extracts.

PROTEIN HYDROLYSATES: PEPTONES, ETC.

A bewildering variety of hydrolysates of various cheap sources of protein, including meat, casein and soya protein, is available. The nomenclature has some logic: peptones are pepsin hydrolysates, and tryptones or tryptoses are trypsin hydrolysates. Pancreatic digests and acid hydrolysates are also common. In general, there are few real differences between them other than in the support of growth of very fastidious organisms, in which case the presence of inhibitors may be a more important factor than the absence of nutrients. Nevertheless, for reliability and for some degree of reproducibility between batches, the better-known products of the large manufacturers may be best for exacting work. A standard, rich basic medium in molecular biology is Luria-Bertani (LB) broth or agar, made from tryptone (1 per cent w/v), yeast extract (0.5 per cent w/v) and NaCl (1 per cent w/v), adjusted to pH 7.5 with sodium hydroxide. For reproducibility between labs, although other media would equally be suitable for many procedures, the scientific community is substantially helped if researchers adhere to standard methodology for common procedures.

SUPPLEMENTS

Specialized media for particular fastidious organisms can be formulated by addition of various additional ingredients or supplements to a peptone base. These are usually either complex, rich sources of a variety of undefined growth factors, such as yeast extract, malt extract, whole blood or serum, or specific factors known to be needed for particular organisms, which may be added either to complex media or to defined media. Such factors may include vitamins such as thiamine or biotin, cofactors such as cocarboxylase, or metabolic precursors such as purines or pyrimidines. They may also include special salts such as iron salts at higher than trace levels, calcium, magnesium or other divalent cations, or reducing agents, commonly thioglycollate or cysteine, to aid pre-reduction and enhance reduction potential in anaerobic media. Sometimes specific ingredients may be incompatible – phosphate added as a buffer salt may precipitate with divalent cations. The concentration of the latter should then be reduced and added slowly with stirring to the complete medium.

Complete media

It is not the intention here to provide detailed recipes for the great variety of complete media which have been described, particularly for the isolation and identification of organisms in clinical material and in public health and hygiene, and the food and water industries. A few examples of types of media, their components and uses, are given in Table 4. Recommended sources of information are listed at the end of the book. General principles however are that for trouble-free routine use, media should be prepared reasonably frequently and using uniform methods and ingredients. This is particularly important for fastidious organisms, and one or two simple precautions may help to prevent variability in quality of media. Where undefined growth factors are needed, supplements (e.g. yeast extract) may best be sterilized by filtration rather than by autoclaving which may destroy labile components. When heat is used, it should be uniformly applied: autoclave times should be minimal (but not too short to kill spores), and media should not be repeatedly re-autoclaved. Agar may lose its gelling strength through repeated autoclaving or melting – although it is often convenient to prepare large batches, allow it to set for storage and melt and pour plates when needed. When adding labile supplements to basic media, care should be taken to adjust temperatures first – ideally by use of water baths. Blood, serum or, for example, filter-sterilized supplements, should be pre-warmed and added at the lowest temperature permitted by the melting point of the agar – provided plates are poured quickly and the gel does not start to set before pouring.

Preparation of media

AGAR PLATES

Pouring of plates is a routine task but one which should be done properly if problems are not to arise. Agar is best prepared in conical flasks with cotton-wool bungs covered in foil or paper, and a convenient size not to tire the wrist or become too hot to hold is a

Table 4 Examples of types of media and their composition.

Type	*Components*	*Uses*	*Selected organisms*
Minimal	Carbon/energy source, e.g. glucose. Nitrogen source (NH_4Cl or amino acid); P as PO_4^{2-}; S as SO_4^{2-}; Na, K, Ca, Ng, Mn, Zn, Fe, Cl^-; vitamins (e.g. biotin); agar if solid	Non-fastidious organisms, e.g. *E. coli.* With supplements, isolation of auxotrophic mutants	*E. coli*; *B. subtilis*; *S. cerevisiae*; *P. aeruginosa*
Nutrient broth or agar	NaCl, soluble extract of meat	Routine growth of non-fastidious organisms	As above and any common laboratory organisms
LB broth or agar	Good quality peptone (e.g. tryptone); yeast extract; NaCl	Routine growth of organisms in molecular biology	As above
Blood agar	Blood agar base (beef-heart infusion, tryptose, sodium chloride and agar) or nutrient agar, and sterile defibrinated horse blood added at 55°C	Many common host-associated organisms and more fastidious species	As above and, for example, streptococci
Mac Conkey agar	Peptone, sodium taurocholate, lactose, phenol red	A typical selective medium, especially for lactose-fermenting enteric organisms	*E. coli*, other enterobacteria, pseudomonads,, etc.
Thiogly-collote broth	Casein hydrolysate, yeast extract, glucose, cysteine, sodium thioglycollate, serum supplement for more fastidious species	Growth of anaerobes. Sterility testing	Spirochaetes, other fastidious strict anaerobes

1-litre flask with 500 ml of agar. It should be cooled to a few degrees above setting point, normally about 55°C, so that condensation to the petri dish as it cools will not be excessive. If the job is urgent, a useful trick is to cool the flask under a hot tap with gentle swirling – tap water is usually supplied at 50–60°C. Supplements should be added with flaming of flask necks, and again mixed gently – many media froth readily and this will lead to bubbles on plates which are difficult to disperse and will be a hindrance to bacterial culture.

Because lids will be removed for some seconds during pouring, the operation should be done in a clean environment, either a room where the air has been static for some hours or in a sterile air cabinet. Plates should be poured to about 20 ml of agar per plate, just enough to cover the bottom of the dish with slight movement to distribute it. Care should be taken not to allow agar to slop up the side of the dish – this may provide a route for entry of contaminants from moisture in and around the lid and for them to multiply. If a few bubbles do form in the agar, they can be dispersed by flaming the surface of the agar briefly with a bunsen burner (blue flame) before it sets – but do not melt the side of the dish! Plates may usefully be stacked in piles of three to five, which can be lifted by holding the lid of each in turn with those above it to pour each plate.

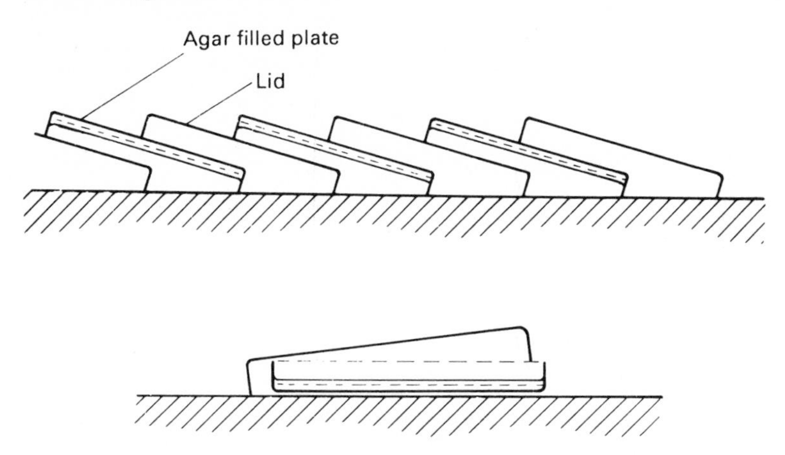

Fig. 21 Drying of agar plates by inversion in a domino pattern to allow free ventilation.

The mutual warming of the stack again cuts down on condensation by reducing the number of cool lids for moisture to condense on.

When set, plates usually should be dried. Many bacteria grow best on a fairly dry agar surface, and if individual colonies are wanted or viable colony counts are being done, liquid must be prevented from running across plates and dispersing the multiplying bacteria. Drying can be achieved in a warm environment such as a still incubator, or in a clean air flow, by inverting plates and lids in a 'fallen domino' pattern, or by slightly opening lids (Fig. 21). Either method must be done with careful regard to sterility. An alternative with triple-vented plates is simply to leave them stacked overnight in a warm room with circulating air – the vents allow enough ventilation to dry the agar significantly. Overnight incubation also allows growth of contaminants so that contaminated plates can be discarded, and indeed plates or other media may be pre-incubated for this purpose.

Prepared plates should be stored in polyethylene bags, conveniently those they were supplied in, until used, to prevent desiccation.

LIQUID MEDIA

Liquid media are most conveniently sterilized in the containers in which they are to be stored and used. This is easily achieved with bottles, but other glassware (e.g. flasks), may be too precious to tie up in storage, and may be filled from bottles when needed. In bottles larger than 20 ml volume, the pressure generated by autoclaving liquids in sealed containers will shatter them, so caps should be left loose and tightened when removed from the autoclave. This is especially useful for anaerobic media – the autoclaving will drive off air from solution. If caps are tightened before cooling, a partial vacuum will form and the audible breaking of this vacuum when the bottle is opened will be a check on continued sterility during storage. Note that any presence of liquid around screw caps, especially from water-bath immersion, will almost certainly lead to contamination of the contents upon opening. Liquid media in flasks with cotton-wool bungs should not be stored for long enough periods for significant evaporation to occur. Close-fitting foil caps may help prevent this.

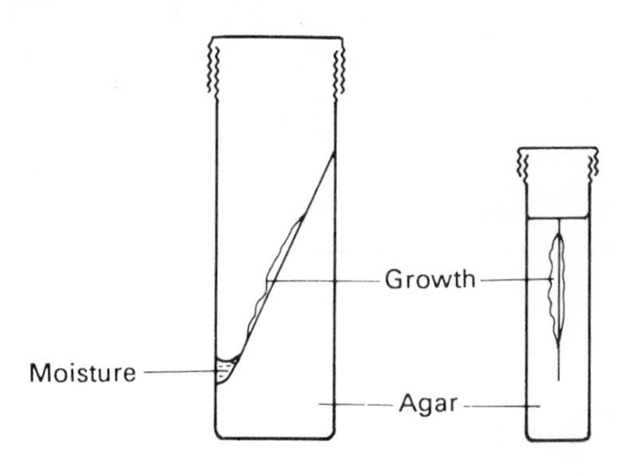

Fig. 22 Agar slope and stab cultures, useful for medium- to long-term storage of cultures.

OTHER FORMS OF MEDIA

Semi-solid media or sloppy agar are mentioned above. Prevention of convection enables an oxygen gradient to be established from the surface downwards into the sterile medium from which air has been driven during autoclaving, so that an optimum can be found for the growth of microaerophilic organisms. Agar may also be prepared in tubes for 'stab' culture (inoculated with a straight wire stabbed into the medium from above), again providing a lowered oxygen level in the depth of the agar. Stabs may also – because of the depth of agar and diffusion of nutrients towards and metabolites away from the stab, combined with protection from dehydration – provide an excellent environment for long-term storage of cultures. Another good system for this is the slope – agar medium is sterilized in screw-cap bottles which are then cooled at an angle so that the agar sets on a slope, with thick agar at the bottom. They are not dried. Again, depth of agar for provision of nutrients and accumulation of moisture at the bottom of the slope provide good conditions for long-term storage of cultures inoculated onto the surface by loop. These storage methods are illustrated in Fig. 22.

Incubation and growth

Gas-phase incubators are the standard facility in medical microbiology, where a temperature of 37°C is commonly used, since organisms of medical importance have to be able to grow at this temperature. In practice, it is also a very useful temperature for many teaching and research applications when standard organisms such as *E. coli, Staphylococcus*, or *Bacillus subtilis* are being studied in basic biological research. Commonly, such organisms are incubated overnight, so the hour or so which may be required for thermal equilibration in air, and some temperature fluctuation when doors are opened, are not important. When a variety of temperatures are required however, or incubation times are short, water baths or heating blocks are more convenient.

Some slower growing organisms may require days of incubation, and special attention to aseptic technique is necessary if contaminants are not to be common and to overtake the organisms being studied. Gas-phase incubation is convenient for longer term cultures.

Fastidious organisms may require special gaseous environments. *Neisseria gonorrhoeae* (gonococci) will only grow at raised CO_2 levels which can be provided in a candle extinction jar, a sealed metal (heatproof!) container in which a nightlight candle is allowed to be extinguished after sealing as the oxygen is used up by the flame and CO_2 produced. *Campylobacter* often require both raised CO_2 and reduced, but not absent, O_2 levels – they are microaerophilic – but candle extinction jars are often not satisfactory and special gas-generating packs for use in anaerobic jars are available, similar to those used for anaerobic organisms (see Chapter 8).

Sometimes growth of fastidious organisms will fail, or be very poor, for no very obvious reason. If the medium is at fault, there may be characteristic growth only at points of heavy inoculation on agar plates. A wise precaution in any case is to check the very basic requirements of incubation temperature, medium pH, and gaseous environment before looking for more obscure reasons for failure.

Acquisition, maintenance and storage of cultures

From time to time every laboratory will need to acquire cultures, whether for teaching, quality control or standardization purposes or research work. Standard strains of many species can be obtained for a small charge from culture collections, some of which are listed with names of other suppliers on p. 155. It is quite safe even for inexperienced students to handle cultures of, for example, *E. coli*, provided they are authentic laboratory strains known to lack virulence factors. Cultures of many useful saprophytic species, for teaching purposes, for example, can be obtained by inoculating puddle or pond water, unlikely to be contaminated by sewage or other human products, onto ordinary nutrient media. Some experience will be needed however for unequivocal identification of, for example, *Bacillus subtilis* or *Pseudomonas aeruginosa*. No attempt should be made to isolate organisms from human sources without appropriate experience or advice in the identification of resulting isolates, which might include potential or actual pathogens.

MAINTENANCE AND STORAGE OF CULTURES

For day-to-day manipulation of cultures in experiments, stocks will often need to be maintained for a few days by subculturing, usually on solid media. There is a dilemma in this procedure, which should be considered although there is no universal rule dictating the best procedure. Should one subculture single, isolated colonies, several colonies or a broad sweep of bacterial growth? There is a risk in taking a single colony, that it will not represent the majority present in the culture, having perhaps a significant mutation which will affect its properties. On the other hand, the progeny of a single colony will be genetically homogeneous, so will behave uniformly in experimental procedures and provided they have the desired genotype, this will be advantageous. If a broad sweep of culture is taken to represent the majority population present, there may be a risk of picking up a contaminant, unnoticed in an area of confluent growth. Perhaps inoculation from a selection of isolated colonies, all appearing identical and true representatives of the cultures, is the best compromise.

These are the arguments for and against single colony picking – each case must be considered on its merits!

It is generally inadvisable to maintain cultures in the longer term simply by subculturing repeatedly onto fresh medium every few days. Apart from the expense of materials used, there is a risk that the character of the culture may change by mutation, or that a contaminant may be introduced accidentally and, if it looks very similar to the original culture, may not be noticed and may replace the original. Some contaminants may outgrow the original culture, perhaps by spreading over an entire agar plate in one night – not impossible if, for example, a *Proteus* strain was the culprit.

The best policy then is to maintain a stock culture for medium-term storage, and to inoculate cultures for use from such a stock. Convenient methods for establishing stock cultures include the use of agar stabs and slopes (see Chapter 8). If left undisturbed, LB agar stabs of, for example *E. coli*, in small bottles with sealed screw caps, stored at room temperature in the dark, should remain viable for up to 12 months. For longer term storage, samples of cultures should be either frozen or freeze-dried. Most bacterial cultures are best preserved by snap-freezing, unlike cultured animal cells which require controlled freezing. Freezing in media supplemented with serum (10 per cent v/v) and glycerol or dimethylsulphoxide (10 per cent v/v), will preserve cultures for months, years or indefinitely, depending on the conditions. At $-20°C$, cultures may remain viable for several months, but for added security a temperature of $-70°C$ is preferable. For indefinite storage, freezing in liquid nitrogen can be used.

Freeze-dried cultures of many species will also keep indefinitely at refrigerator temperatures and are quite stable at ambient temperatures. Organisms should again be suspended in a cryo-protectant as for freezing, and small volumes (e.g. 0.2 ml), dispensed into small soda-glass tubes, about $10\,cm \times 5\,mm$. After snap-freezing, the tubes are rapidly evacuated in a freeze-drier, and left for several hours for all residual moisture to be removed. They may then be heat-sealed in a gas flame, still under vacuum for the best long-term preservation.

CHAPTER 5

Sterility and aseptic technique

Sterility and how to achieve it

Sterility is the complete absence of viable vegetative cells or spores of microorganisms. It is not identical with antisepsis or disinfection, in which potentially harmful organisms only, or the majority of them, are killed. Also, sterility has a slightly different meaning bacteriologically than it has in virology in the context of, for example, filtration – media may be bacteriologically sterile after filtration but may still contain viruses. Sterility is an absolute term, and because of the potential of even one microorganism to replicate rapidly to enormous numbers, there is no place either semantically or in practice for something which is 'almost' or 'partially' sterile!

Several different methods are routinely used for sterilization, and some of them are an integral part of the routine of operations designed to keep things sterile during manipulation – the so-called aseptic (*not* antiseptic!) technique.

DRY HEAT

This may be intense and rapid, as in flaming (see below), or at a lower temperature for a prolonged period. Dry heat sterilization in an oven is of limited value, since bacterial and fungal spores are more resistant to dry than to moist heat: macromolecular structure dependent on hydrogen bonding is the main target of heat sterilization, and hydrogen bonding is less important for molecular stability in the dehydrated state. A minimum period of heating of 45 min at 160°C is recommended. In addition, there may be substantial warming and cooling periods. Thus the materials and equipment are out of action for several hours. Nevertheless, this is a convenient method for some items which are preferably kept dry and have a good thermal stability and conductivity for even heating, such as glassware (especially pipettes) or dissecting instruments. It is not suitable for items with insulating properties such as folded textiles.

MOIST HEAT

This is the most versatile method available, and is based on the use of the autoclave. Note that boiling, while sufficient in many cases for antisepsis (e.g. of dentists' instruments which can be freed of all potentially harmful pathogens in a steamer), will not guarantee bacteriological sterility. The autoclave is quite complex in its principles and practice of operation, and both should be well understood for its safe and effective use. The basic principle is that pure steam at pressures above atmospheric rises to higher temperatures than boiling point, and in use therefore articles must be exposed to pressurized, pure steam to attain the required temperature. The most important problem which may arise is the presence of air mixed with the steam. This will contribute its own partial pressure to the total pressure in the system, and if the resulting partial pressure of the steam is significantly reduced, its temperature at a given total pressure will be reduced correspondingly. Therefore every effort must be made to exclude air from the system. The simplest model of autoclave resembles a domestic pressure cooker (Fig. 23). A heavily constructed container has a wide lid clamped by a heavy duty device to resist the considerable force exerted on its large surface area by a pressure of up to 1 atm. As with media bottles, the larger the surface, the greater the force generated by raised pressures, and even quite small pressures may be very destructive if exerted over a large area! A gas or electric element at the bottom heats a water reservoir to produce steam. This permeates round the load, and escapes through a valve at the top of the vessel. It is allowed to escape freely until it is pure steam, which can be tested by dipping a rubber tube fixed to the outlet below the surface of a container of water – if air is present bubbles will rise, while pure steam will condense in the water with a noise like an espresso coffee machine.

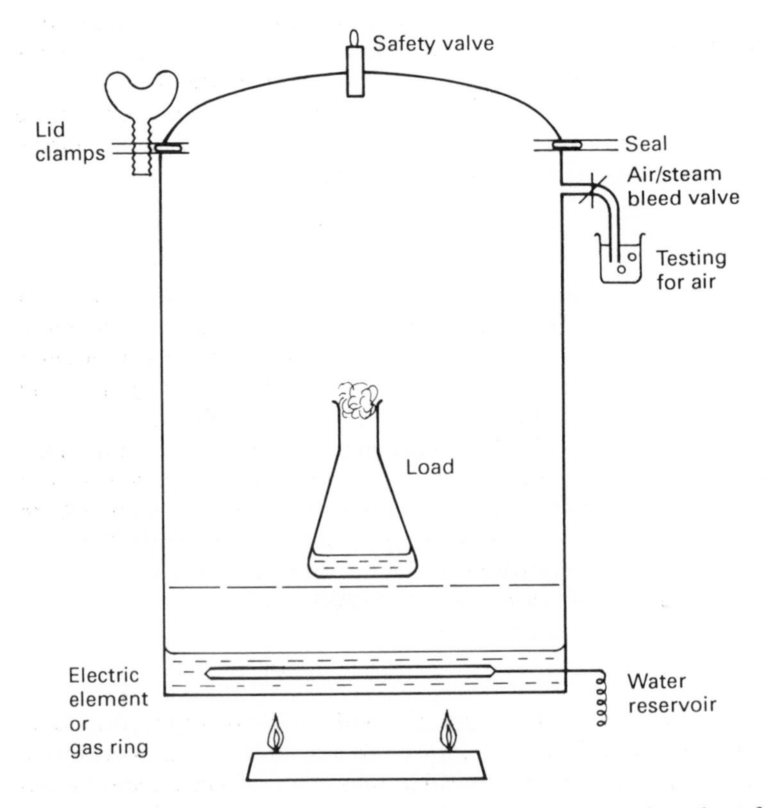

Fig. 23 A simple autoclave. As in a domestic pressure cooker, the safety valve also acts as a pressure regulator.

When the steam has flushed out all the air, the valve is closed and the pressure allowed to rise to the desired value, usually 1 atm (1 bar or 15 lb in^{-2}) above atmospheric which will cause the temperature to rise to 121°C. This value should be held for a minimum of 15 min, assuming all the contents have time to reach this temperature. Large volumes of liquid (> 500 ml) will take longer. After a sufficient time the heat is turned off and the contents allowed to cool and the pressure to come down gradually during cooling. Pressure must not be released prematurely or liquids will boil over, bungs may be dislodged and containers may even burst. Also the contents, if liquid, must be allowed to cool substantially before removal. They may for a while be superheated above boiling point, and movement may trigger sudden boiling and expulsion of contents at great risk to the operator.

Large free-standing autoclaves, which may be powered from a steam main, are often programmed to operate automatically. Valves close to raise the pressure when effluent steam and condensate reaches a required temperature. However, it is still necessary to understand their operation when dealing with atypical loads, and for safety reasons if there is any fault with their operation. Such machines may require monitoring to ensure their effective working. Coloured indicators (autoclave tape which changes colour after effective treatment) and bacterial spore preparations which can be checked for killing, may be used. A thermometer connected to an exterior gauge which can be inserted in the load may also be available. Precautions in loading may also be necessary with large autoclaves: even though steam is an excellent vehicle for heat transmission due to its carriage of latent heat (hence in part the superior efficiency of the autoclave over the dry oven), items should not be packed too tightly to allow the circulation of steam, and large, deep containers with lids should be avoided – lids should be loose fitting so that steam can penetrate throughout. All dry, capped glassware should be autoclaved with loose lids, both to allow heat penetration and so that sterilization is by moist and not dry heat.

IRRADIATION

Irradiation is a useful general method of sterilization, but not of much practical value in the laboratory. The most effective form is high-energy gamma-radiation, but this is only practicable on a large-scale manufacturing basis where the expense of a radiation source such as ^{60}Co, and its careful control and monitoring, can be justified. Many pre-packed sterile plastics are sterilized by this method. In the laboratory, the most commonly used irradiation method is by exposure to ultraviolet radiation, often overnight, of workplaces and benches, especially sterile air cabinets. This is actually of limited value, and should be regarded more as a means of disinfection than of sterilization. This is because it takes a finite time to kill all organisms and spores in a given volume of air, and when doors have been opened and closed and air has moved during entry of the operator, the surrounding air will no longer be sterile. In addition, of course, the radiation cannot be continued in the presence of the operator. Bench surfaces are far more effectively disinfected chemically: corners, crevices, and dirty and shaded areas are often not effectively reached by ultraviolet light.

CHEMICAL METHODS

Again, many chemical methods of sterilization are ineffective as methods for preparation of sterile apparatus or, for obvious reasons, media, which have subsequently to support microbial growth. Exceptions are volatile chemicals, and two examples are in general use. Sterilization by alcohol (most commonly ethanol, although isopropanol is also used quite often, for example in clinical work to disinfect skin) is quick and effective in clean conditions, for example for smooth surfaces, instruments and especially for the traditional glass spreader used in viable counting on agar plates (see Chapter 9), in which flaming is used to complete the process and to remove the excess alcohol. Note that ethanol is most effective at about 70 per cent by volume in water (ethanol:water, 70:30).

The second chemical treatment which is highly effective, but again of little use in the laboratory, is exposure to ethylene oxide. This powerful alkylating agent is very effective in a manufacturing environment for sterilizing disposable plastics, but is highly toxic and inflammable and cannot be used except in carefully controlled facilities; being highly volatile it is quickly dispersed from the articles being sterilized.

FILTRATION

The merits of different types of filter are discussed above (p. 18). In general, modern methods rely on membrane filters, predominantly of nitrocellulose and available in various formats. When first introduced on a large scale a couple of decades ago, nitrocellulose filters were generally mounted in re-usable, autoclavable holders which were assembled in the laboratory and autoclaved before use. This is still common practice with larger units, but there are often problems with shrinkage of membranes away from seals if the units are not assembled very carefully, or with imperfect cleaning of the support grids so that traces of dirt may cause damage to the membrane during autoclaving. For this reason, pre-sterilized, disposable units are now commonly used for smaller volumes. Filtration is particularly useful for sterilizing small volumes of media and supplement solutions for addition to media sterilized by autoclaving, and when volumes are small, pressure can usefully be applied with a syringe to filtration units which have the appropriate tapered inlet port. For larger volumes, some arrangement for applying either negative pressure below the filter or positive pressure above, with a vacuum or pressure pump, must be devised. An outlet from the filter which allows aseptic dispensing of the sterile effluent must be arranged. These systems are shown in Fig. 11.

STERILE SUPPLIES PREPARATION

This is an important part of the efficient maintenance of sterility, mainly reliant on the autoclave, or in the case of pipettes and glassware, the oven. Glassware should be clean and dry. This involves soaking in a suitable detergent – several excellent alkaline detergents are sold for the purpose – washing by machine or by thorough brushing, ideally an acid wash in dilute HCl (e.g. 0.1 M) to neutralize charges introduced by alkaline detergents, and several rinses in distilled water. If capped, glassware for autoclaving should have the caps left loose so that steam can penetrate. Bungs for culture flasks, and plugs for pipettes (necessary to help prevent accidental suction of pipette contents into the teat or device, and to filter dirty air entering the pipette), should be of non-absorbent cotton-wool (which is slightly oily and therefore water repellent). Pipettes should be plugged with care, with about 2 cm of cotton-wool (a straightened paper clip is a good tool – Fig. 24), so that plugs do not come adrift in

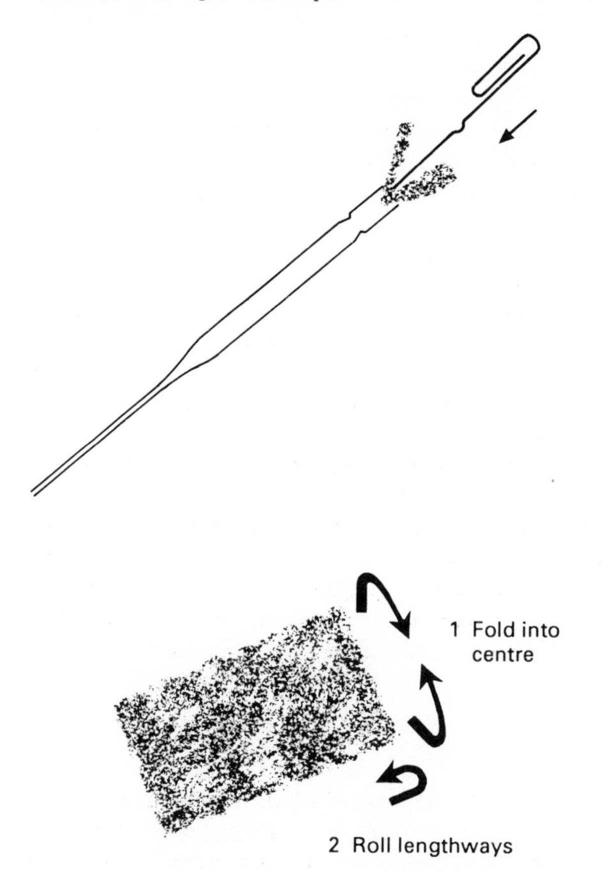

Fig. 24 (Top) Use of a paper clip to plug pipettes with cotton-wool. (Bottom) Plugs for conical flasks, etc. can be formed by smoothing the cotton between thumb and index finger.

use but can be removed for cleaning. Machines driven by compressed air are available for pipette plugging if done on a large scale, although they are not entirely satisfactory and the steady introduction of disposable pipettes may discourage investment in them. Plug removal, if difficult, is best done by tapping the plugged end of the water-filled pipette (stood in a bucket of water to fill) lightly on a hard rubber surface – the column of water will act as an effective ram to push the plug out. Cotton-wool bungs in culture flasks, tubes or bottles (carefully made by rolling or smoothing round a finger (Fig. 24) to retain their shape and not shed excess fibres) should ideally be covered with foil or steam-permeable greaseproof or Kraft paper (strong when wet) secured with Scotch or autoclave tape or a rubber band, so that the neck is protected from contamination after autoclaving. Paper and tape, or single-use purpose-made paper bags, can be used for wrapping and autoclaving a variety of items such as dissecting instruments, polypropylene Eppendorf tubes or micropipette tips. The last can also be autoclaved in purpose-made plastic holders with holes for the tips and lids, which double as storage and dispensing racks.

Aseptic technique

The object of aseptic technique is to prevent any form of contact between the sterile growth medium or apparatus, or the pure culture, which is being handled and the various environmental sources of contaminating organisms. For the beginner, there is a curious mental block which has to be overcome before constant awareness of possible contact with contaminants becomes automatic. Without this awareness, which comes to most of us only with practice, it is surprisingly easy to flame a pipette to sterilize it, and then put it down on a non-sterile bench while we pick up something else, for example. The sources of contamination we have to keep in mind are any non-sterile surfaces, liquids and the air in which operations are carried out. The air is particularly important not just as the source of perhaps the occasional free-floating fungal or bacterial spore, but as carrier of sizeable particles, liquid droplets or solid dust or dried skin scales, which in turn may be carriers of vegetative organisms or spores. In addition to the problem of training the mind to be aware of all routes for contamination, there is also a mental process of acquisition of confidence that the procedures followed should, and must work, if followed correctly. Sometimes problems of contamination can be chronic and soul-destroying, especially if they occur at the start of a project and we have yet to see the evidence that success can be achieved. Often we then seek explanations or even excuses, blaming the weather, the season, the sterile supplies procedures, or the air conditioning, but most often there is a faulty technical procedure which can be corrected.

FLAMING

Flaming is the classical procedure for rapid sterilization of loops, pipettes, and bottle and flask necks. It is often not done very well: surfaces being flamed will need more than a quick pass through a flame to take up enough heat to raise their temperature high enough to kill all organisms. A finger can be passed rapidly through a flame without burning it! Therefore except for the loop and pasteur pipette tip which have low thermal capacities, flaming needs to be done for several seconds – long enough to raise the temperature of surfaces well above 100°C – if organisms present in, for example, secretions from the fingers left on glass surfaces, are to be killed. Bottle and flask necks should be rotated slowly in the flame for 5 s or so to achieve this. Flaming is also useful to remove stray cotton-wool fibres from plugged pipettes before use with teats or pipetting devices, so that they do not destroy the air-tight seal necessary for suction.

CAP, PLUG AND PIPETTE MANIPULATION

For basic procedures of liquid transfer and inoculation, manual dexterity in removing and holding caps and plugs is essential. The most reliable technique is to use the little fingers of each hand to unscrew and hold caps and plugs, leaving the other digits free for manipulations. This is not easy to do at first. Untrained little fingers have little dexterity or strength (as any typist knows), and practice is required to attain these. Also difficult to control is the pasteur pipette with rubber teat: the best arrangement is to hold the pipette between the index and middle fingers (leaving the third and fourth fingers entirely free to hold bottle caps), and operate the teat by pressure applied with

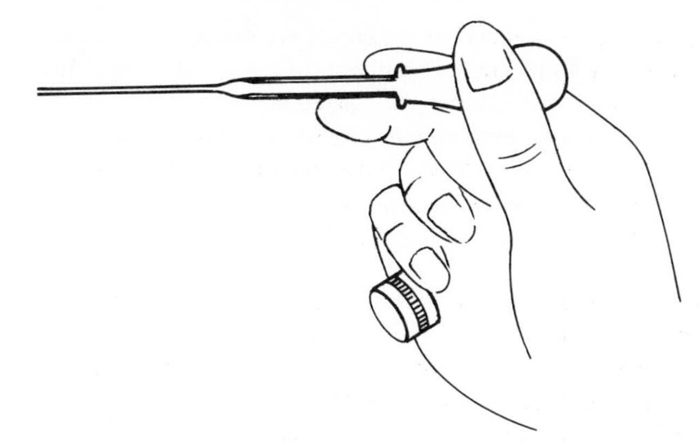

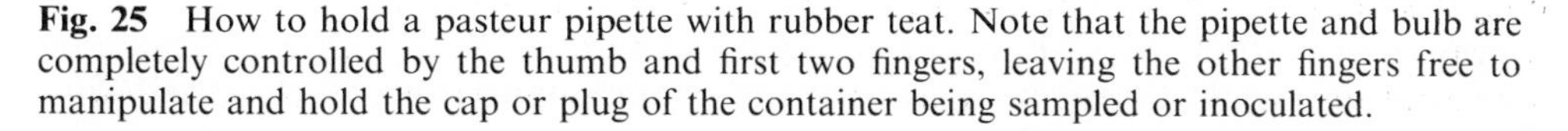

Fig. 25 How to hold a pasteur pipette with rubber teat. Note that the pipette and bulb are completely controlled by the thumb and first two fingers, leaving the other fingers free to manipulate and hold the cap or plug of the container being sampled or inoculated.

the thumb against the top of the index finger (Fig. 25). Again, coordination is difficult for the beginner. It should, however, be routine to take a pasteur pipette in the right hand (assuming a right-handed operator), remove the cap from a bottle held in the left hand using the little finger of the right hand, withdraw liquid, replace the cap (retaining the pipette in the right hand), take another bottle in the left hand, remove the cap and expel the liquid into it, and replace the cap, flaming the bottle necks with the left hand after opening and before closing.

One additional hazard in dirty air (i.e. not in a laminar flow cabinet) is airborne contamination. Usually most airborne contaminants are borne on particles of significant size, and on the open bench the operator is the most likely source of these. Breathing, and especially coughing and sneezing and talking, all lead to some production of aerosols (particles of liquid too small to fall rapidly through the air) and more importantly larger droplets, of respiratory secretions and saliva, which contain viable organisms. The skin, hair and clothing also shed particles, especially dry flakes of dead skin, which also carry organisms. Both types of particle will be shed by the careless operator around the working area, and often they will fall quite rapidly through the air. For this reason another important aspect of aseptic technique is to preclude the deposition of particulate material into open bottle necks etc., by keeping them in a near-vertical plane as much as possible: open containers are held on their sides. The same applies to petri dishes: if opened for inoculation or inspection they should be held vertically as much as possible. Also in storage or during incubation of petri dishes, attention to the vertical route of particulate contamination is important. Dust may fall either from the environment into the edge of the lid if the plate is upside down, or from the atmosphere within the lid onto the agar surface if the plate is right way up. Usually plates are incubated upside down (with the agar at the top), partly in order to prevent any moisture present from running across the agar, but it should then be remembered that contaminants may gather around the inside edge of the lid, and if there is any moisture there it may become contaminated. In such circumstances the plate and lid must remain upside down for as long as the culture is required.

Transfer of small volumes of liquids is most safely achieved by pipetting. In some situations, however, the extra time (while containers are open to the air) and manipulation involved in pipetting can be a greater risk of contamination than the alternative of pouring liquids. Provided it is certain that necks of bottles or flasks are clean or can be flamed adequately, and if it is possible to pour without spilling liquid or wetting the mouth of the receiving vessel, pouring the total contents of one vessel into another is reliable. The little fingers of both hands can be used to hold the cap or bung of the container in the other hand.

Special risks of contamination may arise with screw caps if the threads become wetted. This may occur in a water bath if liquids are being thawed or warmed before use: if they are accidentally submerged they should be discarded even if screwed up tight, since it is impossible to sterilize threads by flaming or otherwise before unscrewing, and capillary action will then contaminate the inside (wetted also with the contents) as soon as unscrewing commences. Also if screw threads are wetted with medium or culture from within the bottle, capillarity will carry it down to the non-sterile bottom of the thread and environmental contaminants will gain a hold and work their way up to the top of the thread, again being impossible to kill effectively.

SPECIAL PROBLEMS IN ASEPTIC TECHNIQUE: LAMINAR FLOW

For many purposes adequate sterility can be achieved by strict application of the methods and precautions outlined above, even on the open bench. In some circumstances, however, extra precautions are necessary because of the nature of the materials or equipment used or the incubation conditions. Very humid conditions are often difficult to work with: some fastidious organisms grow best on moist (undried) plates, and moisture may accumulate around petri dish lids, forming a fertile ground for growth of contaminants; high humidity in jars used for anaerobic or microaerophilic conditions may also encourage such problems. Long-term cultures of slow-growing organisms in rich media may also be susceptible to overgrowth by small numbers of contaminants which might not be a problem in overnight cultures of rapidly multiplying organisms inoculated at high levels. Tissue cultures or others which are predominantly done in plastic containers cannot be handled with the same element of flaming in aseptic technique. In such cases the laminar flow cabinet may help to provide the extra level of protection against airborne contaminants needed for maintenance of sterility, but there should be no relaxation of the other elements of technique described above. The cabinet should be used to maximum effect by following a few simple rules:

- The fan should be switched on at least 10 min before use so that any dirty air and dust which may have entered and settled in the working area or the downstream side of the filter will be flushed away.
- The bench surface should be disinfected and cleaned with a suitable detergent or alcohol preparation and allowed to dry.
- The bench should be as free of objects as possible, particularly upstream of the manipulation area, so that the air will be truly laminar: any turbulence induced by bulky objects may be vigorous enough to suck in dirty air from outside (Fig. 6).

CHAPTER 6

Safety

There are three main aspects to the anticipation and prevention of hazards arising from the actually or potentially infectious nature of microorganisms. First, precautions can be taken to prevent contact with all or any cultured microorganisms handled in the laboratory. Second, an understanding of the basic processes of infection and the potential of different organisms to cause disease will help the informed operator to anticipate and prevent unnecessary hazards. Third, a special category of potentially hazardous organism is now being encountered increasingly: the artificially engineered recombinant created by genetic manipulation, for which there are control measures for handling and use to prevent their dissemination in the environment.

Prevention of contact with laboratory microorganisms

The aim of safe handling of laboratory cultures is to prevent whatever possible contact can be foreseen between viable microorganisms and the operator. It should be assumed that all organisms or cultures, even if known (in theory) to be harmless, are potentially hazardous. This is because firstly, although many organisms are harmless to people in normal health, they may not be to those who have immunological deficiencies, and secondly, cultures may be contaminated and may in fact comprise, or be mixed with, pathogenic organisms.

Contact with organisms may be by direct handling or exposure to cultures or contaminated material, or by indirect contact through contaminated glassware or other discarded materials, or by airborne exposure.

BASIC HANDLING TECHNIQUES

This should be performed with care and with every precaution to prevent contact. First, laboratory routine should be designed to minimize risks. Suitable laboratory coats should be worn; if clinical material or known pathogens are being handled, coats should be of an approved, full-length pattern made of an absorbent, autoclavable material with side or back rather than front fastening, high throat, and elasticated wrists to minimize the risk that spilt liquids will penetrate. In case they do, and use of disinfectants may be necessary, a shower should be available. Again, in laboratories where clinical material or pathogens are handled, or for some categories of gentic manipulation work, separate sinks with elbow-operated taps for hand washing should be located near the exit. Eating, drinking and application of make-up in laboratories should be prohibited, and mouth pipetting should be totally forbidden. Ideally, as noted in Chapter 2, laboratories should be designed with reading and writing areas separate from the working bench.

The working area itself can be guarded against small spills spreading rapidly by the use of an absorbent cloth, moistened with *disinfectant*, spread on the bench. Purpose-made, disposable cloths impregnated with a disinfectant can be bought. Disinfectant in a liquid form should be available, both in discard jars and in a readily available source such as a wash-bottle. For general purposes, the most useful are the amphipathic detergent type, such as Tego, or glutaraldehyde-based, e.g. Cidex; these are non-corrosive and safe in use, and the first also cleans if used to wipe down benches. Alternatives are dilute hypochlorite and phenolics; for the merits of each, see below. The working area should also be free from clutter and well-lit, at the correct working height, and with adequate room to move freely.

Manipulations of cultures may in themselves create hazards if done incorrectly. The most insidious hazard, assuming that all reasonable precautions are taken to prevent careless spillage of culture onto surfaces which are subsequently touched or handled, is probably generation of *aerosols*. Formation of these is quite difficult to avoid, and since they are essentially invisible, the operator who is not informed of the risk will not be aware of their presence. Experiments with liquids stained with fluorescent dyes and illuminated with ultraviolet light, with high-speed photography to record the track of droplets and aerosols, have shown that almost all manipulations which lead to the breaking of a liquid surface cause the formation of greater or lesser amounts of aerosol. An aerosol is essentially colloidal – particles are small enough to remain in suspension in the air for substantial periods – so they will be carried around by air currents. They therefore pose a significant hazard if formed from suspensions of dangerous pathogens. Procedures which form aerosols should be conducted carefully. Some aerosol formation accompanies the use of wire loops and pipettes, especially if bubbles are broken in the course of liquid handling. More extensive aerosols form during any kind of liquid mixing – homogenization, ultrasonication, or even vortex mixing – and will be released when caps are removed afterwards. Centrifuges are a potential source of aerosol formation, especially if liquid is spilt within then so that the rotor and air movement can disperse it in small droplets. If organisms are able to survive in aerosol form and present a significant hazard (see below – 'Processes of infectious disease'), they should be handled, at least for procedures likely to produce extensive aerosols, in a safety cabinet as described below.

DISCARD OF CULTURES AND CONTAMINATED MATERIALS

Discard procedures carry a large part of the potential risk associated with handling of microbial cultures. The containers needed are described in Chapter 2. All unwanted cultures and supernatants, etc. should be placed in discard cans or buckets for autoclaving. Petri dishes containing agar cultures, and other disposable plastics should be discarded separately, or if mixed with glassware, separated in autoclavable plastic bags. Pipettes should be discarded into containers so that they are completely immersed in a suitable disinfectant. There is no hard and fast rule about suitability and recommended concentration of disinfectant for this purpose. The laboratory supervisor should consider the local circumstances, categories of organism used, and the type of use (e.g. nature of solutions – high or low protein concentrations, volumes, frequency and quantity of material discarded) in drawing up appropriate procedures. The merits and disadvantages of different types of disinfectant are outlined below. Hypochlorite, phenolics and amphipathic detergents are all potentially useful, but hypochlorite should not be used if jars are subsequently to be autoclaved because it may corrode the autoclave. It should be used at a concentration between 1 and 10 per cent by volume (1000–10000 ppm of available chlorine) of the concentrated solutions such as Chloros (10 per cent sodium hypochlorite) which are available commercially. The concentration needed will depend on the degree of soiling by organic matter, especially protein, likely to be encountered. Phenolics such as hycolin can be used in discard jars at a concentration of 1 per cent, or more where high concentrations of organic matter are expected. Discard jars for small items such as used microscope slides or micropipette tips (which may also be discarded into autoclavable bags) will also be needed, and for contaminated sharp items such as needles and scalpel blades, disposable incineration containers are recommended, such as the Cin-bin supplied by Labco.

PREVENTION OF AIRBORNE EXPOSURE

This is the main function of the *safety cabinet*, although it does have the additional advantage that small spills are also contained within a defined space, and the hazardous nature of the organisms handled is clear to all concerned so that unauthorized workers will stay away. The three basic types of cabinet are described in Chapter 2, and are illustrated in Figs 26–28. The use of the gloved class III cabinet, for organisms requiring total containment, is highly specialized and should only be contemplated by highly trained personnel. Basic rules for use of the other types of cabinet are however fairly straightforward. The cabinets do have limitations: even when working perfectly they cannot contain splashes of liquid or very heavy aerosols

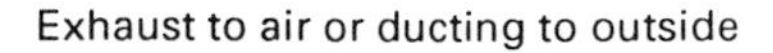

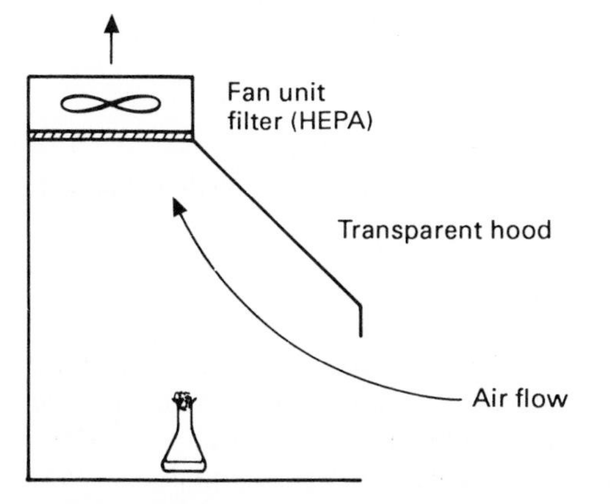

Fig. 26 A class I safety cabinet. Note that the air flow is away from the operator, quite opposite to that in the laminar flow cabinet (see Fig. 6). The air flow will, however, carry airborne environmental contaminants towards the work.

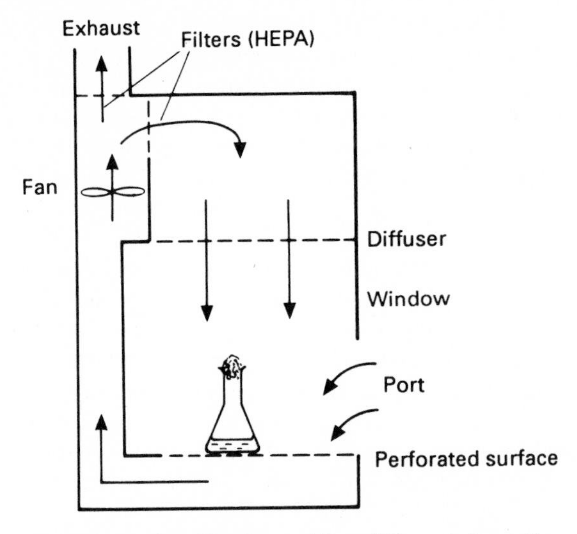

Fig. 27 A class II safety cabinet. Dirty air from the environment makes minimal contact with the work, being augmented by a stream of clean air from above, thus enhancing the sterility of the working area.

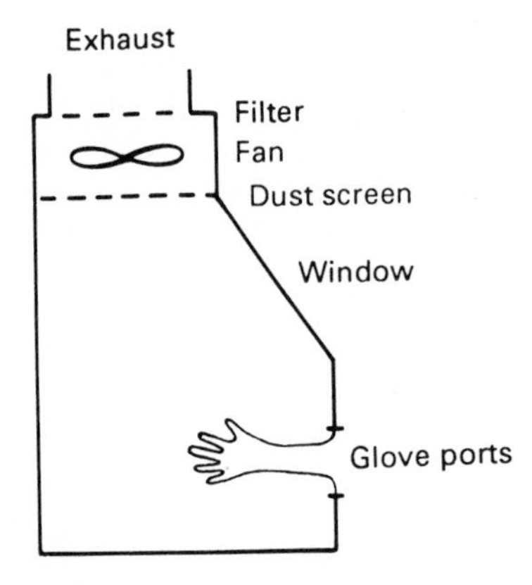

Fig. 28 A class III safety cabinet. Normally the filter exhaust would be ducted to the outside air. The working area is totally enclosed, with access via glove ports.

produced with turbulence of the surrounding air, for example by a centrifuge. Air turbulence both within and around the cabinet will reduce its efficiency drastically. There should not be large objects close to the working aperture, and the interior of the cabinet generally should be uncluttered. The aim is to achieve a near-laminar flow into the cabinet, as with the outward flow of a sterile air cabinet. Movement of air outside the cabinet should be prevented by restricting passage of people, and siting it away from doors and windows.

DISINFECTANTS

Disinfectants all have advantages and disadvantages and there is no one 'best' choice for all laboratories.

Hypochlorite While highly effective when fresh and relatively non-toxic to man, hypochlorite is corrosive, especially to stainless steel and aluminium which are normally regarded as stable to corrosion in moist environments. It is easily neutralized by organic matter. Its biggest disadvantage is the hazard of handling concentrated solutions in which it is supplied. These are strongly oxidizing, and will give off toxic levels of chlorine if mixed with acids or certain salt solutions, as well as being harmful (e.g. to fabrics or wood). Even in dilute solutions of working strength some damage to the latter can occur with long-term exposure, and of course splashes can disfigure clothing by bleaching.

Phenolics These are effective, non-corrosive and generally more resistant to neutralization by organic matter. They may however have an unpleasant odour and may leave permanent films on glassware, especially if it is autoclaved after disinfection.

Glutaraldehyde This is also quite resistant to neutralization by organic matter, and is fully water-soluble and non-corrosive. It is however more harmful to man, as is *formaldehyde* (normally available as a 40 per cent aqueous solution, formalin) which is mainly used in the gaseous state for fumigation. The latter is used for cleaning up after serious accidents, or for sterilizing working areas or safety cabinets before maintenance work. Addition of 7.5 g of potassium permanganate to 35 ml of formalin will result in release of gaseous formaldehyde, in sufficient quantities to fumigate a safety cabinet which should be closed and sealed after the release of the gas. If the cabinet is not exhausted to the atmosphere, it may be necessary to leave the room overnight to clear the fumes after this treatment!

Detergents, including cationics such as cetrimide, and amphipathic detergents such as Tego, are also quite effective against vegetative organisms, are quite resistant to neutralization by organic matter, and are non-corrosive and harmless to man.

Processes of infectious disease

In order for an infectious disease process to exhibit the full spectrum of effects including damaging host cells or tissues, several conditions must be fulfilled. In general, the healthy host will only be susceptible to known, unequivocal pathogens. There are however also environmental organisms, as well as organisms of the normal flora associated with our own internal and external body surfaces, which can in some circumstances cause disease, usually due to some weakness or 'compromise' in our host defence systems. These organisms are often called 'opportunistic pathogens', or 'opportunists', and may include organisms which are being studied for reasons quite unrelated to their pathological properties. Some categories of environmental organism can never be pathogens – for example if their optimum growth temperature is well below that of the body. Even these organisms may occasionally be harmful, for example if they produce toxins during processes of food spoilage.

The processes necessary for disease causation include some, or all, of the following:

- colonization by the pathogen of host surfaces
- entry of the pathogen into the host
- multiplication of the pathogen within host cells or tissues
- resistance of the pathogen to the killing or neutralizing effects of host defence mechanisms
- ability of the pathogen to damage the host

Each of these may usefully be considered in reaching an understanding of the basic processes of infectious disease.

COLONIZATION OF HOST SURFACES

Many pathogens, as well as harmless commensal members of the natural flora of the alimentary tract, are able to colonize host surfaces by specific colonization mechanisms. These often include the ability to stick to host cells or tissues, mediated, for example, by surface structures such as fimbriae commonly found on the surface of Gram-negative bacteria which may undergo highly specific ligand–receptor interactions with host-cell surface components such as glycoproteins. Because of the specific nature of such interactions, and their role as essential steps in pathogenesis, pathogens are often quite host-specific; for example, enteric pathogens of domestic animals may not be pathogenic for man. In some cases, however, they are: food poisoning organisms such as *Salmonella* species are often acquired from domestic animals. Colonization mechanisms are especially important for organisms which have pathogenic attributes which enable them to harm the host while located at the mucosal surface. Sites of potential colonization are not limited to the alimentary tract. We have to remember the respiratory tract, eye, genito-urinary tract and indeed the skin, in relation to possible accidental colonization by laboratory organisms. Often columnar epithelium is more susceptible than squamous, although squamous cells may have their own indigenous microflora, for example many species of streptococci in the mouth specifically colonize particular areas of squamous epithelium such as the cheeks. The notorious *Streptococcus mutans* colonizes tooth surfaces very efficiently, and has a well-understood role in tooth decay.

ENTRY INTO THE HOST

Organisms which have a mode of pathogenesis which entails multiplication within the host need to enter the tissues. If colonizing a mucous epithelium, this will often be by a process, not fully understood, which appears to resemble phagocytosis by host epithelial cells. Infection may, after initial penetration, remain essentially confined to the epithelial layers of tissue, as with for example dysentery bacilli in the gut or *Neisseria gonorrhoeae* infection of the genito-urinary tract. Non-mucosal infections often involve entry through wounds or quite minor abrasions in the skin. Saprophytic or environmental organisms with opportunistic potential may enter the host this way, or normal inhabitants of the skin itself, such as staphylococci, may do so, although unbroken skin may be quite resistant to penetration. Thus it is important that open wounds or abrasions should be covered with waterproof plaster during experimental work. Especially likely mechanisms for accidental penetration of organisms are puncture wounds, which may be from broken glass – particularly pasteur or graduated pipettes which quite often break in inexperienced hands when teats or pipetting devices are fitted to them – or from the purpose-made hypodermic needle! The latter is often used to inoculate or withdraw infected material from animals, eggs, etc. and sooner or later it is likely to penetrate the skin even in experienced hands. A few simple rules such as never to re-sheath needles will substantially reduce the risk.

MULTIPLICATION WITHIN THE HOST

In most cases, organisms need to multiply within the host as a step in pathogenesis. There may be particular nutritional or physiological requirements. For example, some of the more notorious environmental opportunists such as *Clostridium tetani* are strict anaerobes, and are therefore unable to multiply in the relatively well-oxygenated environment of living tissues. They do thrive however in anaerobic conditions which may develop in deep, dirty puncture wounds or in the presence of dead tissue or cells. Hence the need for such sites to be cleaned and maintained in a healthy state after injury.

Another nutritional requirement of all living cells is for iron; although it is only needed at very low concentrations, its availability in healthy living tissues and in the blood stream is also very low, due to the low solubility of iron at physiological pH and also to the presence of highly efficient iron-chelating proteins in the blood and mucosal secretions. Some bacteria, especially pathogenic species, have equally, or more, efficient iron-scavenging mechanisms of their own, which enable them to acquire iron even from the host iron-binding proteins and thus multiply within, for example, the blood stream. For such reasons, organisms from clinical sources are more likely to have pathogenic potential and should generally be handled with more care than environmental isolates.

RESISTANCE TO HOST DEFENCE MECHANISMS

The human body is protected very efficiently from the vast majority of microbial challenges. Every time we brush our teeth, suffer minor cuts and abrasions, ingest food or water contaminated with environmental organisms, or breathe the air close ot others who are talking, coughing or sneezing, we inevitably come into close contact with potentially infective organisms. The first two processes definitely introduce living bacteria into the blood stream. And yet many of us live for year after year without the slightest sign of active infectious processes. This is because our defence systems cope very adequately with small numbers of organisms of no particular pathogenic potential. Phagocytes in the blood stream and tissues are able to ingest many organisms even without the help of serum proteins, antibodies or complement, which may aid the process in difficult cases.

Antibodies are proteins produced by the cells of the immune system which specifically recognize and bind to foreign antigens, which may be proteins, carbohydrates or lipids present on the surface of microorganisms. We do not necessarily have to have been exposed to specific organisms before in order to have some 'natural' antibody activity against them – often non-pathogenic organisms possess common antigenic molecular structures, for example in their cell walls, which antibodies will recognize because they cross-react (share three-dimensional structural shape) with those from other organisms. Once antibodies bind to the surface of bacterial cells they may either enhance the process of phagocytosis by which microorganisms are ingested by phagocytes and killed by intracellular digestion, or cooperate with complement proteins to cause direct lysis, particularly of Gram-negative bacteria.

Another common and important lytic mechanism arises from the action of lysozyme, a ubiquitous host enzyme present in mucosal and exocrine secretions, for example. Pathogens have quite specific mechanisms for avoiding many of these host defences which are so effective against environmental organisms. For example, pathogens may have polysaccharide capsules which are antigenically unlike the surfaces of common environmental bacteria, and which are therefore unlikely to be recognized by the pre-formed antibodies of the normal individual. Capsules may also prevent efficient phagocytosis or lytic killing by humoral factors. Alternatively, the surface of a pathogen, the critical site in determining the outcome of the interaction between host defence systems and invading microorganisms, may be structured in such a way that host defences are ineffective. Antigens may resemble those of host cells so that they are not recognized by the host immune system as foreign. Microbial surfaces may specifically acquire and bind host proteins, again making the microbial surface 'host-like'. Many such mechanisms of avoidance of host defences are understood in some detail.

DAMAGE TO THE HOST

Most pathogens have well-defined mechanisms by which the host is damaged or its normal physiological processes are perturbed. The best-known examples are of toxin-mediated damage, as in the classical infections of diphtheria, tetanus and cholera. The first two diseases entail distribution of toxins around the whole body and their action on a variety of target cells (although the 'lethal hit' may be on particularly vulnerable essential life-support systems such as the heart or respiratory systems). On the other hand, cholera toxin acts specifically on the intestinal epithelium, at the same site as colonization by the organisms, to cause a disturbance of electrolyte and water transport across the intestinal epithelium, leading to net loss of water from the blood stream to the gut lumen and rapid, potentially lethal dehydration of the host. Less overtly pathogenic organisms, for example many opportunists which do not have powerful host-damaging mechanisms, may have to reach high numbers in the body before damage is done, but nevertheless if they multiply unchecked the outcome may be serious. A frequent example is bacteraemia or septicaemia by Gram-negative bacteria of various kinds, often opportunists, in debilitated patients: 'endotoxin' (lipopolysaccharide), a normal constituent of the cell walls of Gram-negatives, has complex toxic and haemodynamic effects when present in the blood stream, and will lead to a state of 'septic shock' which may be rapidly lethal.

The above examples show that pathogenicity is often a multi-factorial property of microorganisms. Deletion of any one of a number of factors from a potent pathogen may effectively disable it: a cholera *Vibrio* which either could not adhere to the intestinal epithelium, or could not produce an effective toxin, would not be highly pathogenic. Similarly, expression of just one potential virulence factor, for example possession by a commensal organism of the gut of an adhesion mechanism, or expression of a toxin by a recombinant organism artificially derived in the laboratory, will not usually create a virulent pathogen. Nevertheless, as soon as circumstances

Table 5 Pathogenic processes and virulence factors in selected bacterial infections.

Process	*Disease*	*Factor*	*Organisms*
Colonization/ adhesion	Diarrhoea	Fimbriae	*E. coli*
	Gonorrhoea	Fimbriae	*N. gonorrhoeae*
	Dental caries	Glucan	*Streptococcus mutans*
	Streptococcal sore throat	M protein	Group A streptococci
Entry into host	Bacillary dysentery	?	*Shigella dysenteriae*
Growth within host	Urinary-tract infection	Iron-induced outer-membrane proteins	*E. coli*
Avoidance of host defences	Meningitis	Sialic acid capsule	Group B *Neisseria meningitidis*
	Lobar pneumonia	Capsular polysaccharides	Encapsulated pneumococci
	Antigenic modulation	Fimbrial antigens	*Neisseria gonorrhoeae*
		Variable outer-membrane proteins	*Borrelia hermsii*
Damage to the host	Diarrhoea	Enterotoxin	*Vibrio cholerae*; *E. coli*
	Diphtheria	Toxin	*Corynebacterium diphtheriae*
	Tuberculosis	Antigenic components	*Mycobacterium tuberculosis*

obviate the need for a particular virulence factor, infection may result from unexpected sources. For example, host debility may lessen the efficiency of phagocytic cells so that an organism without an anti-phagocytic capsule but perhaps with the ability to acquire iron in the blood stream, may multiply there, causing a potentially dangerous septicaemia. An understanding of these processes may help laboratory workers to anticipate hazardous procedures or situations. A summary with some examples of pathogenic processes and virulence factors is given in Table 5.

Regulation of work with microorganisms

All work with microorganisms is now ultimately subject to regulation by the *Health and Safety at Work Act (1974)*. An innovative provision of this legislation was that both employers and employees have elements of duty to ensure safety in the workplace. Thus all workers must be informed and trained to an appropriate level in the hazards they may encounter. Clearly therefore, laboratories must be adequately supervised, with a designated safety officer, and appropriate training must be available. In turn, employees have a legal duty to adhere to procedures laid down for the laboratory, which must be readily available for consultation.

Detailed regulations for safe handling of microorganisms apply only to clinical laboratories, to the handling of a relatively small number of designated pathogens, and to genetic manipulation *in vitro* and the handling of genetically manipulated organisms.

CLINICAL LABORATORIES AND WORK WITH DESIGNATED PATHOGENS

The principal regulatory document is the report of the Advisory Committee on Dangerous Pathogens (ACDP), *Categorisation of Pathogens according to Hazard and Categories of Containment (1984)*. This document lists all microorganisms considered to be potentially pathogenic and categorizes them according to the level of risk perceived. It also defines the levels of containment appropriate for handling pathogens of each category, in terms of facilities and laboratory conditions which must be met.

Four categories of organisms are defined:

Group 1 These are organisms which are most unlikely to cause disease in man, and this includes all organisms not listed in the other groups.

Group 2 These are organisms capable of causing disease, which may be a hazard in the laboratory but rarely cause laboratory infection and are unlikely to spread in the community, and are easily controlled by prophylaxis or treatment. This group includes the vast majority of routinely encountered pathogens and many species commonly used in the research laboratory, including the ubiquitous *Escherichia coli.*

Group 3 These are organisms capable of causing severe disease, which present a serious hazard in the laboratory and may have the potential to spread in the community. Only a dozen or so bacterial genera are represented. They are usually treatable in some way or prophylaxis is available.

Group 4 These organisms are similar in hazard to Group 3, but effective prophylaxis or treatment is not available. Group 4 organisms are all viruses.
For further information see the ACDP report and other sources listed on p. 151.

Containment is defined at four levels, each appropriate to organisms of the corresponding group.

Level 1 This is for harmless organisms, essentially involves no containment measures other than observance of normal hygiene – no eating or drinking or other obviously foolish practices – and the laboratory must have washing facilities, effective disinfectants and discard procedures, etc.

Level 2 These laboratories must be large enough for the workers accommodated ($24\,m^3$ per person), easily cleaned, used only for microbiological work and with restricted access. Suitable coats (see above, p. 49) must be worn. There must be access to an autoclave for discard of cultures and materials, for which safe arrangements must be made. Procedures likely to generate extensive aerosols must be performed in such a way as to contain the aerosol, normally in a class I safety cabinet. Clinical laboratories must contain a class I safety cabinet.

Level 3 This includes all the above, and in addition the air must be continually drawn into the room and expelled to the atmosphere through a HEPA air filter capable of removing all infectious particles. All work must be done within the laboratory, so all equipment needed must be available and dedicated for the purpose. Entry must be restricted. Manipulations must be done in a class I or III safety cabinet, and gloves must be worn.

Level 4 In addition, this level entails many regulations about entry and leaving the laboratory through an airlock, use of double-ended autoclaves to remove materials, etc. and is a very highly specialized facility.

It is clear from the above that there are relatively few restrictions on work with known isolates of strictly harmless organisms, which may be handled safely in virtually any laboratory provided the basic rules of hygiene and good practice are observed. Thus work in schools, for example, is perfectly feasible. For work of any serious nature, which will almost certainly sooner or later involve organisms in category Group 2, a reasonably well-organized laboratory will be needed, following the general principles of good practice outlined at the beginning of this chapter. Provided it is not used to handle clinical isolates, the restrictions will not severely limit procedures used even at an advanced research level. Work with the more dangerous classes of pathogen will not be done outside a medical context with full awareness of pathogenic hazards and their containment.

Regulation of genetic manipulation

Shortly after the development of methods for artificially recombining DNA molecules from various sources in deliberate ways in the laboratory, scientists began to ponder the potential hazard of the creation of new combinations of genetic material which might not occur in nature. Suppose for example that a gene for highly lethal diphtheria or botulinum toxin was transferred to, and expressed in, a strain of *Escherichia coli* which was highly transmissible in man and able to colonize the alimentary tract. Would there be an epidemic of fatal disease as the organism colonized the population? Or suppose the potential existed for the transfer of these or other genes to human tissues? The possibilities are endless (once the dogma, that DNA transfer and its expression between species was essentially impossible, had broken down) and rather frightening. Public opinion was alerted about these dangers and a need for regulation of such work was clear.

As a result, in the late 1970s, committees of experts in various countries were formed, to devise schemes for the control of such work and the containment of recombinant DNA and the organisms harbouring it. A dilemma soon became apparent: was all artificial manipulation of DNA equally dangerous, or were some types of experiment essentially safe, and others potentially extremely hazardous? In the UK, a rather elegant scheme of risk assessment was devised, based on logical premises, although initially untestable for its efficacy, which would have to be monitored through experience. Along with the categorization of risk for various procedures, again there were containment levels defined rather similarly to those for work with pathogens so that organisms harbouring recombinant DNA would not be released into the environment. The regulation of this type of work was initially overseen by a group known as GMAG – the Genetic Manipulation Advisory Group. It has now been replaced by the Advisory Committee on Genetic Manipulation, set up by the Health and Safety Commission and administered by the Health and Safety Executive. Regulatory documents were originally in the form of *GMAG Notes*, most of which have been superseded by revised ACGM notes, obtainable from HM Stationery Offices.

RISK ASSESSMENT

Risk assessment attempts to base the need for containment on the assignment of numerical values to definable risks associated with different categories of work. Estimates of numerical values for probabilities of hazardous events occurring are made in three areas:

- that recombinant DNA will be released and persons exposed to it – 'access'
- that it will have the potential to be harmful to exposed persons – 'damage'
- that gene products will actually be expressed from the DNA in its recombinant form – 'expression'

(1) *The access factor:* This will depend on the probability that accidental exposure to a culture of a recombinant organism will result in colonization of the exposed person. In the case of *E. coli*, a natural inhabitant of the gut, this probability will be high (for practical purposes with a value of (1) if the *E. coli* strain is a 'wild-type' isolate with no handicaps in establishing itself in the gut. If the strain is deliberately selected to be 'disabled' – for example, it may lack all known factors which enable it to adhere to mucosal surfaces and resist host defences – we may assume there is a lower probability that it will establish an infection, and we can estimate that probability by assuming, for example, that it would have to mutate to wild type before being able to colonize, and we can assign a probability of perhaps 10^{-3} that significant access to man would occur. In practice, even if a strain was unlikely to colonize the gut of an exposed person, it might very well transfer recombinant DNA to wild-type strains already present in the gut. Therefore, an important additional limitation on access which is often used is based on the inability of certain strains of *E. coli* to transfer their DNA to others, and such factors will further reduce the probability of access occurring. This is known as biological (as opposed to physical) containment.

(2) *The damage factor:* This is the probability that if access is achieved, the gene products encoded by DNA will actually have the potential to harm the host significantly. At one extreme would be, for example, a defined sequence of DNA encoding a known protein, present and abundant in every individual in the same form and with no known pharmacological or other potentially harmful action, perhaps a structural protein such as actin. This might have a probability of being harmful as low as 10^{-9}, based on the possibility that the gene might mutate to an abnormal and potentially harmful form in its recombinant state. Unknown sequences, such as random genomic fragments, would potentially be more harmful – if from man, for example, they might encode a hormone which could have harmful effects if expressed at high levels in inappropriate tissues. The probability that any one recombinant, among tens of thousands present in a gene bank representative of the entire genome, will express such a protein will, however, be < 1, and it can be assigned a value. At the other extreme on the damage spectrum, we can imagine DNA known to encode a toxin highly lethal for man. This would have the highest probability of damaging the exposed host, and would carry a risk factor of 1.

(3) *The expression factor:* This relates to the probability of recombinant sequences being expressed at biologically significant levels. It is known that the most abundant proteins in bacteria may be present at levels as high as 10^6 copies per cell, whereas the least abundant may only be at a few copies per cell and some are not expressed at all under some circumstances. If the genetic manipulation is deliberately designed to select organisms expressing known genes at the highest possible levels, the expression factor

Table 6 Risk assessment factors and appropriate containment for genetic manipulation of pathogens.

Combined risk assessment factor	*Containment level*	*Main features in addition to those at lower levels*
Less than 10^{-15}	Level 1 – good microbiological practice	Proper well-designed laboratory; protective clothing; no eating or drinking; adequate hygiene
$\leqslant 10^{-12}$–$> 10^{-15}$	Level 2	Separate laboratory; dedicated hand-washing facilities; containment of extensive aerosols; appropriate discard system; restricted access
$\leqslant 10^{-9}$–$> 10^{-11}$	Level 3	Separate equipment for all procedures; negative air pressure with filtration to outside. All work done within safety cabinets.

may have a value of 1. On the other hand if known sequences are deliberately selected so that expression is impossible – for example, a gene is in truncated form – they will be essentially harmless and may have a very low risk factor of perhaps 10^{-12}, again reflecting the probability of mutation or further recombination to a more harmful form.

From the risk factors in the above categories, a combined risk factor can be derived (the simple product of the three), which represents the risk of the procedure overall. For various levels of overall risk containment levels are laid down which parallel those for dangerous pathogens. If the combined risk factor is $< 10^{-15}$ or less, essentially no containment is necessary other than the basic good laboratory practices of hygiene and disposal of containment Level 1 for non-pathogens outlined above. For factors of 10^{-12} or less, but more than 10^{-15}, containment at Level 2 for pathogens is required, and for factors of 10^{-4} or less, but more than 10^{-12}, Level 3 for pathogens must be observed. Containment above this level is very rarely required. These levels are summarized in Table 6.

OTHER ASPECTS OF CONTAINMENT

Several years of practical experience of genetic manipulation work has indicated that the hazards may not be as great as originally feared, and there has been some relaxation of regulations in several countries. A complete revision is under way, with some conformity expected in Europe as the issue is considered at a European Community level. Two aspects however do remain of considerable concern: (1) large-scale use of genetically manipulated organisms, and (2) their planned release into the environment. The latter is particularly fraught with difficulties due to its concern to the Green movement, and further developments can be expected during the next few years.

CHAPTER 7

Growth of microorganisms – theoretical aspects

Whatever the aim of work with microorganisms, the first step will be to grow suitable cultures for study. The practicalities will be very dependent on the organisms and the type of system used (see Chapter 8), but the theoretical framework of microbial growth applies universally. To obtain optimum yields of the most appropriate material, this framework must be familiar to the experimenter.

Exponential growth

As indicated in the Introduction, balanced growth of microorganisms in closed systems, assuming no concurrent death, leads to an exponential increase in number: the rate of increase in number in the population is proportional to the number of microorganisms present. Rate of increase and population size are related by a constant:

rate of increase in cells with time $= k \times$ number of cells present

This constant k defines the rate of exponential growth, hence it is named the *growth rate constant*, sometimes designated μ. The above equation can be expressed mathematically as a differential equation:

$$\frac{dN}{dt} = kN$$

where N is the number of cells in the population at any time t.

Upon integration:

$$\ln N_t - \ln N_0 = k(t - t_0) \tag{1}$$

where t is the time when the number of cells is N_t, and t_0 is an earlier time when the number of cells was N_0.

In terms of $\log_{10}$,

$$\log_{10} N_t - \log_{10} N_0 = k/2.303(t - t_0) \tag{2}$$

If this equation is rearranged, we have an expression for derivation of the growth rate constant k during exponential growth at a constant rate:

$$k = \frac{2.303(\log_{10} N_t - \log_{10} N_0)}{t - t_0} \tag{3}$$

The growth rate constant k has units of h^{-1}.

We shall return to the determination of the value of k below, after considering the 'growth curve' of a so-called batch culture, from start to finish – only a quite small part of which fulfils the conditions of exponential growth described by these equations.

We may ask how the growth rate constant is related to directly observable parameters of culture growth – the most obvious one being the time it actually takes for one cell to complete a whole cycle of growth and division. This is called the *generation time*, and is obviously equivalent to the time it takes for the number of cells in the population to double (each cell has become two cells), i.e. the doubling time. Equation (1) indicates how the relationship can be derived. If we consider the case where N_t is just twice the value of N_0, this represents the exact doubling of numbers in the population. Therefore, in this case, $t - t_0$ equals the mean doubling time or generation time, often called g. We can rearrange equation (1):

$$k = \frac{\ln N_t - \ln N_0}{t - t_0}$$

and substitute g for $t - t_0$, and $2N_0$ for N_t:

$$k = \frac{\ln 2N_0 - \ln N_0}{g} \quad \text{or} \quad k = \frac{\ln 2}{g} = \frac{0.693}{g}$$

It is important to remember that the units of time in these calculations are always hours, whereas we tend to measure growth in practical experiments, where cultures may be growing rapidly in the laboratory, in minutes.

The growth curve

The classical bacterial growth curve (not to be called a growth cycle, to avoid confusion with the cell division cycle – see below) is obtained from the monitoring of cell numbers in a batch culture, i.e. a fixed volume of liquid medium. After inoculation with a small number of organisms from a pure culture, incubation continues (ideally with shaking to provide uniform conditions for every cell) for a long period so that growth is initiated, continues exponentially as long as there are no restrictions of nutrient supply or build-up of toxic wastes, eventually ceases, and may subsequently be followed by a decrease in numbers as cells begin to die. The curve which describes these events is illustrated in Fig. 29, and its main features are as follows. Note that cell numbers are plotted on a logarithmic scale, so curves must be interpreted with caution – they do not always represent what they appear to! Note also that classical batch culture conditions would not normally be found in nature. The graph shown here would be typical for our prototype *E. coli*, growing in pure culture in a reasonably rich medium with shaking.

LAG PHASE

After inoculation of the culture, there is a period during which no increase, or a very slow rate of increase, of cell numbers with time occurs. This lag phase represents a

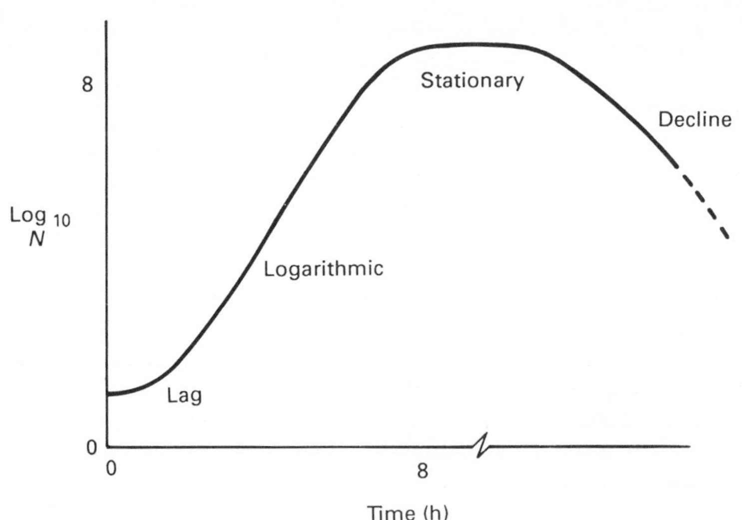

Fig. 29 A typical batch culture growth curve. Since the curve is plotted in a semi-logarithmic manner, the straight line increase in number represents the period of exponential growth. The shape of the stationary and decline sectors of the curve will vary substantially depending on the organism and growth conditions.

period when cells are adjusting to the inevitably different growth conditions encountered in the fresh medium, compared with that from they have been obtained. The lag phase can be minimized, for example by producing the inoculum from a culture already growing at low concentration (so that nutrients are not depleted), exponentially, in an identical medium to that used in the experiment. The lag phase is difficult, if not impossible, to eliminate completely, however – cell concentration will inevitably be lower in the newly inoculated medium than in the source culture, and this may affect growth conditions. Note that the lag phase covers the whole of the initial curved part of the 'curve' – due to the logarithmic plotting of cell numbers. Only the straight line part of the graph represents exponential growth (see below).

EXPONENTIAL PHASE

This is often called the logarithmic or 'log' phase of growth, since it is the period when the logarithm of the number of cells increases linearly with time. The slope of the straight line at this part of the curve relates to the growth rate constant (see below): the steeper the curve, the higher the value of the growth rate constant, and the shorter the doubling time. Merely because the process is exponential does not, of course, mean it has to be rapid: some organisms may only divide every 20 h or so, but they will still grow exponentially with a straight line plot at the equivalent of this part of the curve until conditions no longer support unlimited growth. The rate of growth is also affected by temperature, and so the steepness of the slope at this point will reflect this (see below). Exponential growth can only continue as long as all essential nutrients for growth are available in excess, and no waste product has reached an inhibitory level. In rich media, exponential growth may continue longer than in poor media, provided toxic products such as acids do not reach inhibitory levels before nutrients are exhausted. This particular problem may be countered, but only to a limited extent (since solute concentration, and hence osmotic pressure, may become limiting) by the use of buffers. The ultimate limitation on the extent of exponential growth is that by its nature it is an acclerating process and if unchecked, could simply not continue for very long. An organism dividing around three times, to give a tenfold increase, every hour would produce 10^{24} cells in 24 h. If each weighed 10^{-12} g, this would be 10^{12} g, or one million tonnes of biomass. Quite clearly there are limiting factors – whether we like it or not – to the sustainability of exponential microbial growth in batch culture! In normal circumstances, a culture of perhaps 100 ml of medium, starting from a count of 10^3 ml^{-1} could rise to perhaps 10^9 ml^{-1} during exponential growth, and if increasing about tenfold every hour, this would take about 6 h – conveniently within the working day, but considerably less than the usual overnight incubation period. Thus overnight batch cultures of rapidly growing *E. coli* cannot normally be 'log phase'.

STATIONARY PHASE

As soon as the supply of nutrients or the build-up of waste products begins to limit the rate of growth, it will tail off – usually rather rapidly in rich media because the cell numbers are so high that conditions change very rapidly at this time. In poor media, the slow-down may be more gradual, since there may be alternative nutrients, for example to the preferred major energy source, and cell numbers may be low enough for substantial additional growth to continue (see below for a discussion of diauxic growth). Depending on the physiology of the organism and the density of the culture, the stationary phase may be short or prolonged. Some organisms adapt and survive well in a dormant state, whereas others may be quite labile when they are not actively growing. *Neisseria gonorrhoeae*, for example, has a constant enzyme-mediated turnover of peptidoglycan, the structural cell-wall polymer responsible for cell-wall integrity, and if not actively synthesizing new material it will rapidly autolyse: hence the stationary phase will be short.

DECLINE PHASE

The decline phase may comprise a steep or shallow slope, and death, like growth, may be an exponential process giving the straight line shown in Fig. 29, if certain conditions are fulfilled.

Graphical determination of growth rate constant

We can rearrange equation (2) to the following form:

$$\log_{10} N_t = k/2.303\ (t - t_0) + \log_{10} N_0$$

This is an equation of the form

$$y = mx + c$$

which is the classical form for a straight-line graph with the constant m defining the relationship between x and y (i.e. the slope of the line), and c the intercept on the y-axis.

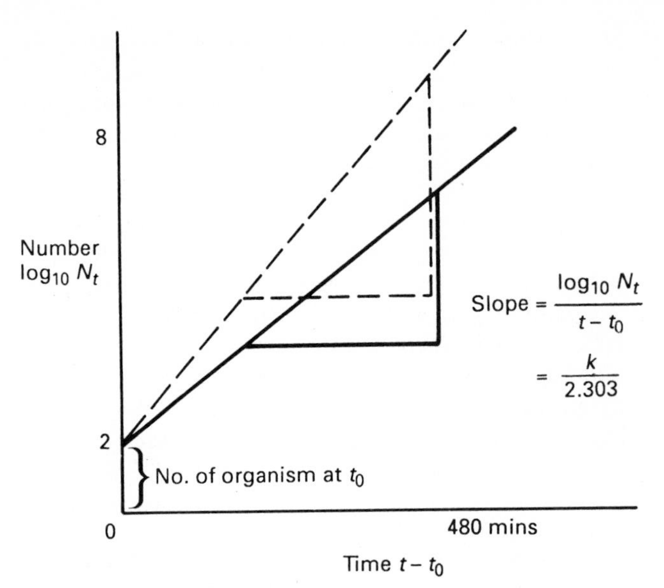

Fig. 30 Determination of growth rate constant from a plot of $\log_{10}$ of the number of organisms present against time (hours). To calculate the slope, the values of $\log_{10} N_t$ and $t - t_0$ are read from the axes at the upper and lower points on the graph, and the lower values subtracted from the upper ones to give the figures used in the equation.

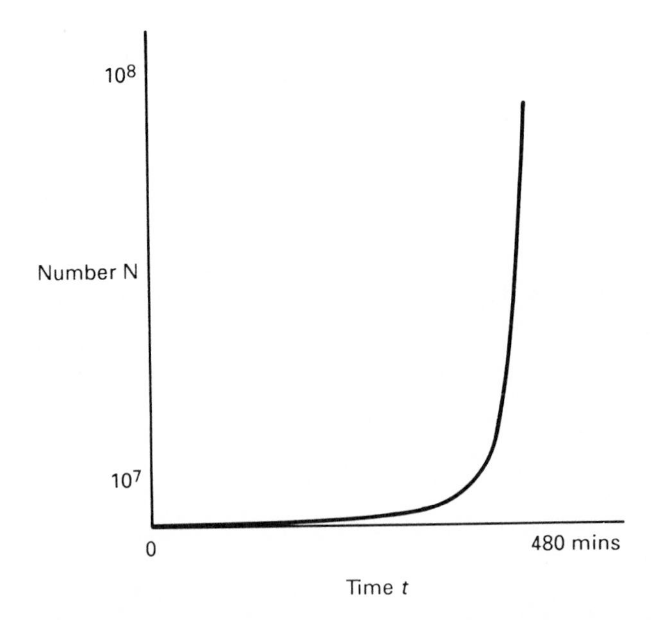

Fig. 31 A representation of the growth curve shown in Fig. 29, but using arithmetic values of N. It can be seen that the curve is much more difficult to analyse mathematically, and the low numbers present at the start of the experiment cannot be represented graphically.

Therefore a plot of $\log_{10} N_t$ (number at time t) against $t-t_0$ (i.e. against time after the start of the experiment) during exponential growth should give a straight line, of slope $k/2.303$, and the intercept on the y-axis will indicate the number of organisms present at time t_0, as shown in Fig. 30. Provided growth is in the exponential phase, this graph can be used to determine the growth rate constant under the conditions of the experiment.

Note that the figures on the y-axis represent cell numbers. Equally they might be obtained from direct readings of some other parameter of growth, such as optical absorbance of the culture. In that case (see the right-hand y-axis) the logarithmic values of absorbances (with numerical values at the lower range being >1.0) would have negative values as shown in Fig. 30. (Those unfamiliar with manipulations of logarithms – since the advent of pocket calculators! – may find this difficult to handle, especially if log tables are used to determine log values of readings: negative logs derived from log tables always have positive values for the mantissa (the part after the decimal point), so the log value of 0.2 would be 'bar 1' 0.3010, or $-1+0.03010=-0.6990$. The calculator should give the correct value!)

For interest, the arithmetic increase in number, during a similar period of exponential growth as shown in Fig. 30, is represented in Fig. 31: it is clear that this is an impractical way to represent microbial growth graphically, as well as being of little use in defining it mathematically.

Constraints on growth rate

TEMPERATURE

We all know that many biological processes, including microbial growth, occur faster at higher temperatures; bread rises more quickly and food decays more rapidly. Often these processes essentially cease at moderately low temperatures, and are severely inhibited and eventually stopped at quite modestly high temperatures. How can these effects be defined mathematically?

The rate of chemical reactions is affected by temperature in a general way as illustrated by the *Arrhenius equation*:

$$\log_{10} v = \frac{-\Delta H}{2.303\ \mathrm{R}T} = C$$

where v is reaction velocity, H is the activation energy of the reaction, R is the universal gas constant, and T is the absolute temperature (K). According to the equation, a linear relationship should then exist between the logarithm of the reaction velocity and the reciprocal of the absolute temperature. We may regard microbial growth as representing the sum of a number of chemical reactions, which should follow the Arrhenius equation in responding to changes in temperature. In fact they only do so over a relatively small temperature range, as shown in Fig. 32, where the logarithm of growth rate constant is plotted rather than that of reaction velocity. Several interesting features of the curve are apparent. There is indeed a linear region to the curve where it follows the Arrhenius equation. Bearing in mind that growth rate is defined by the growth rate constant, it is clear that as temperature rises in this region, the growth rate constant will increase also; the effect on the graphical representation of exponential growth is to make the slope steeper (Fig. 30, dashed line). At low temperatures, the

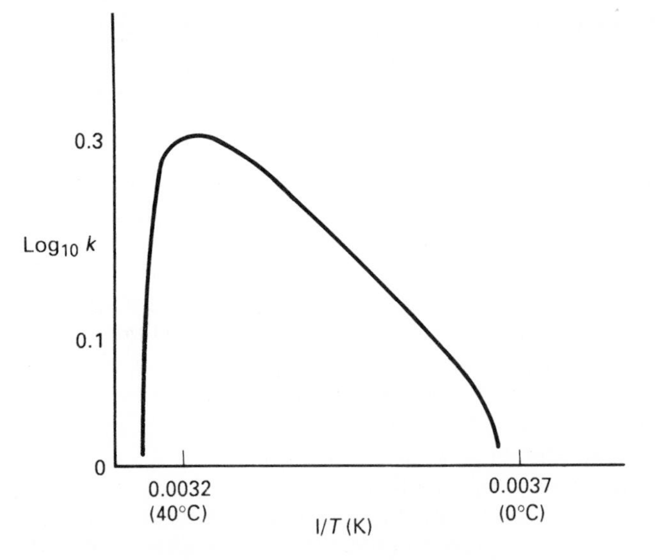

Fig. 32 An Arrhenius plot of the relationship between temperature and growth rate constant. The latter is plotted on a logarithmic scale.

Arrhenius plot departs from linearity: this is because reactions are inhibited by effects other than temperature alone, for example membranes will lose fluidity and cease to function as lipids solidify. At higher temperatures, reaction rate peaks and then suddenly falls off: the reason, or course, is that thermal damage to cell macromolecules (e.g. denaturation of proteins), sets in. This has a dramatic effect on enzyme function, hence the maximum temperature at which growth is supported is very close to the optimum.

Microorganisms are quite diverse in their responses to different temperatures, and they are often classified in three groups – *psychrophiles*, *mesophiles* and *thermophiles*.

Psychrophiles These are adapted to survive and grow at low temperatures, with temperature optima for growth of 0–20°C. They may have increased cold tolerance, for example by containing raised levels of unsaturated fatty acids in membrane lipids, with a correspondingly lower setting temperature so that membranes will be fluid, and functional, at lower temperatures. (Another example of this effect is the lower temperature at which margarines containing unsaturated fatty acids remain soft for spreading!) Such organisms may have a high proportion of rather heat-labile proteins.

Mesophiles These have temperature optima in the range 20–40°C, and include most of the laboratory organisms used on a wide scale, as well as most human pathogens with temperature optima around 37°C. Note that quite modestly raised temperatures, perhaps only 45°C, may be inhibitory or even rapidly lethal for some of these organisms.

Thermophiles These have temperature optima, or at least tolerance, to remarkably high levels. Optima of 50–60°C are commonly found, for example in *Bacillus*

Table 7 Effect of medium composition on growth rate, for an organism such as *E. coli*

Medium	*Approximate growth rate (doubling time, min) at 37°C*
Minimal, amino acid C+N source	60
As above+glucose	35
Defined mixture of amino acids, salts, glucose	25
Undefined; e.g. protein hydrolysate, glucose, salts	20
Nutrient broth	16
Brain–heart infusion, yeast extract	14

stearothermophilus, a well-known thermophile which may be found in the soil, and is capable of raising the temperature in compost or manure heaps well above ambient by its metabolic activity. More extreme thermophiles have been found more recently, and a prominent example is *Thermus aquaticus*, an organism isolated from sea-bed thermal vents where temperatures may be >100°C (which is possible, due to the solute levels and increased pressure under water). Thermophiles have exceptionally stable proteins resistant to thermal denaturation, and their enzymes have been exploited by man for this reason – in biological washing powders, and in a more scientific environment, as thermostable reagents in molecular biology, like the DNA polymerase from *T. aquaticus* which is used in the polymerase chain reaction.

NUTRITION

Availability of nutrients is an important determinant of growth rate. Even those organisms which are able to grow on minimal media do so more slowly than they would do on nutritionally more complete media. We can illustrate this in several ways. Table 7 shows the effect of media of varying complexity on the growth rate constant of *E. coli*. On minimal medium, the doubling time is 1 h; whereas on the most complete, complex media the generation time is <15 min. As more nutrients are added to the minimal medium, the growth rate rises steadily: less work has to be done by the organisms to synthesize all their cellular constituents.

An interesting illustration of the effect of different major nutrients on growth rate can be seen in Fig. 33, which shows the phenomenon known as *diauxic growth*. In a minimal medium, the growth rate of an organism able to multiply rapidly under such conditions, such as again, *E. coli*, is dependent on whatever carbon source is most readily available. In this case such a nutrient might be glucose, and the growth curve shown here has a steep slope initially indicating quite rapid exponential growth until the count of organisms reaches about 10^7 ml. At this point, all available glucose (which would have to be present in the medium at about 0.05 mM) has been utilized, and the organism switches to an alternative carbon source, perhaps another carbohydrate less central to numerous metabolic pathways, or for example an amino acid, which is used less efficiently than glucose and the growth rate falls, as shown by the less steep curve. Note however that the growth remains exponential. Note also that the concentration

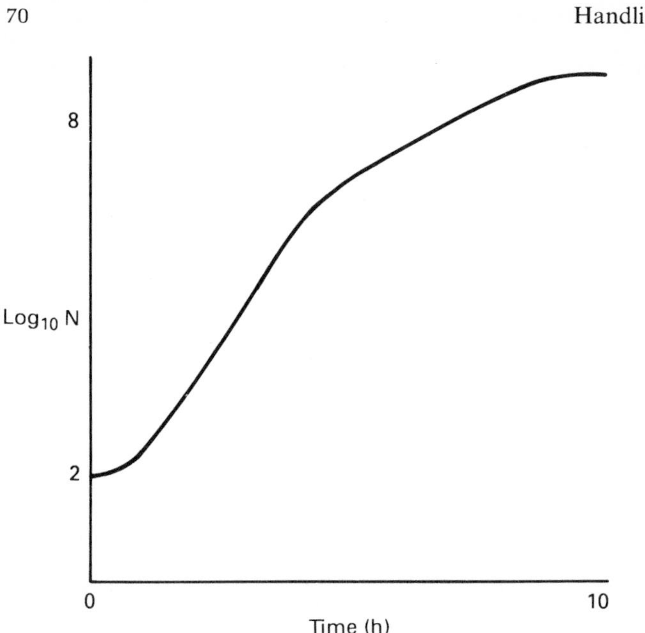

Fig. 33 A diauxic growth curve, obtained when an organism is growing on a minimal medium with two limiting nutrients, for example carbon sources, one of which is used preferentially, with a faster growth rate (reflected by the steeper slope), during the early part of the growth curve.

of glucose used for demonstration of the diauxic growth phenomenon is extremely low – it may seem surprisingly so in the context of levels of perhaps 5–10 mM (e.g. 1 per cent by weight) often used in the formulation of complex growth media. We may ask how growth rate is actually affected by nutrient concentration. By growing organisms at a range of concentrations of an essential nutrient, again preferably in a minimal medium (although an alternative is to use an 'auxotrophic' mutant which is dependent on a particular nutrient in order to grow at all), we can determine what is the effect on growth rate of nutrient limitation. Figure 34 is an example of the type of result seen. It is striking that there is no appreciable effect above a certain rather low value of nutrient concentration in the example shown. At lower levels, there is quite a sharp cut-off point, below which growth is very slow. It is difficult, due to the shape of the curve, to determine exactly the point at which the nutrient is just adequate to support maximal growth, but much easier to determine a point on the curve at the level of half-maximal growth, and this is the value commonly determined from such curves. Those familiar with enzyme kinetics may recognize the shape and form of this graph, and the mathematical treatment which can be used to characterize the biochemical system responsible: the graph reflects the kinetics of an enzyme, in this case the permease enzyme responsible for uptake of glucose by the bacterial cells, which operates at half-maximal efficiency at the remarkably low level of 0.0005 mM glucose. Above this level, glucose is present at more than adequate amounts to be transported efficiently into the cell, to support maximal growth rates – hence the need for a very low glucose concentration in the diauxic growth experiment above. The glucose concentration k_s which allows half-maximal growth rate is equivalent to the Michaelis–Menten

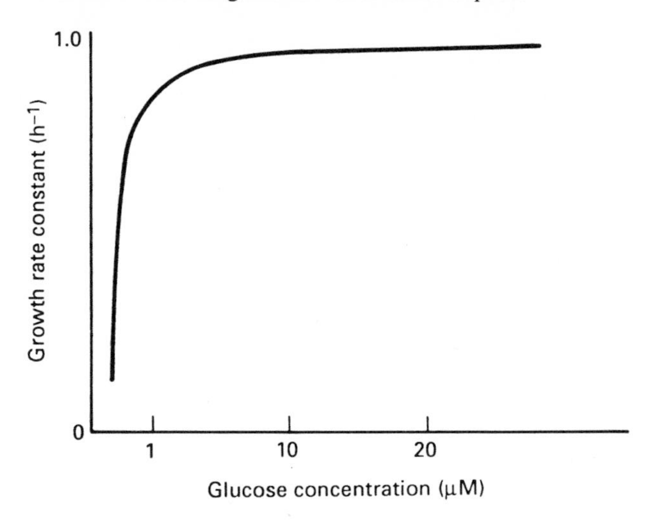

Fig. 34 A typical relationship between growth rate constant and concentration of a limiting nutrient. The curve is hyperbolic, and shows a rapid increase in growth rate around a critical level of nutrient, levelling off at higher concentrations.

constant K_m of the glucose permease enzyme, which thus has a very high affinity for glucose. The equation,

$$k = k_{max}\left(\frac{C}{k_s + C}\right)$$

where k is the growth rate constant, k_{max} is the maximum growth rate constant in the presence of excess nutrient, k_s is the nutrient concentration which supports half the maximum growth rate, and C is the limiting nutrient concentration, describes the hyperbolic curve relating growth rate constant to the level of limiting nutrient. The equation is exactly analogous to that of the kinetics of the permease enzyme. Thus organisms such as *E. coli* are able to grow in environments, perhaps natural waters for example, where nutrient levels may be extremely low.

Finally, in considering the effect of nutrient levels on growth, a useful parameter in physiological studies is the growth yield: the dry weight of cells obtainable (e.g. in mg ml^{-1}) for an equivalent concentration of nutrient (in the same units). This provides information on the relative efficiencies of utilization of different subtrates (e.g. carbon sources) in defined media.

pH

The growth of microorganisms is usually rather sensitive to pH, with reasonable growth rates being achieved only within about 1 pH unit either side of the optimum value. This is illustrated in Fig. 35. Many commonly used laboratory microorganisms have pH optima around netural, although it should be remembered that several well-known species are quite acid-tolerant and indeed thrive in acidic environments – the lactic acid bacteria, lactobacilli and streptococci, the latter notorious for their role in

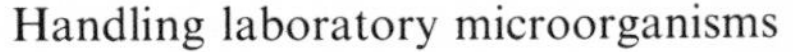

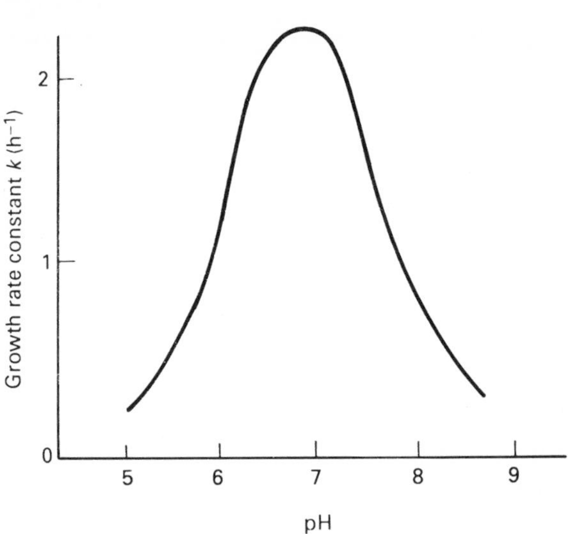

Fig. 35 A typical relationship between growth rate and pH. A fairly narrow range of pH is favourable to growth and such curves tend to be more symmetrical than the temperature–growth rate curve.

dental caries formation, for example, or more benignly in the fermentation of acid foodstuffs such as yoghurts and salami where potential spoilage organisms and pathogens are rapidly killed or at least inhibited from growth. Many fungi, and especially yeasts, are also quite acid tolerant – indeed the isolation media such as Sabouraud's agar used in mycology are often acidic at around pH 5, a useful practical property since environmental bacteria are generally inhibited at this pH.

Some bacteria are tolerant of remarkable extremes of pH, and have been isolated from natural environments at both ends of the pH scale. They include thermoacidophiles, isolated for example from acidic hot springs at $\leqslant$pH 2, and extreme halophiles which may be found in quite strongly alkaline brine lakes. The latter are also very tolerant of, and in some cases actually dependent on, high salt concentrations which may in some cases approach saturation with sodium chloride.

Continuous culture

To this point we have only considered batch cultures. There are however methods for the long-term study of cultures growing, for example, in the exponential phase, without becoming overwhelmed by the resulting biomass! A culture system is set up in which there is a constant turnover, in a fixed volume, of growth medium which is continuously added while simultaneously an equivalent volume of culture is removed at the same rate with its content of microorganisms. Thus a steady-state culture can be maintained for days, weeks or even months (provided the aseptic technique is good enough to keep the culture pure!). A diagrammatic representation of such a system is shown in Fig. 36. The basic needs are simple: a culture vessel, with stirrer to keep the composition of the environment in the vessel uniform; provision for temperature regulation; a port for inoculation and sampling; a system for addition of a constant

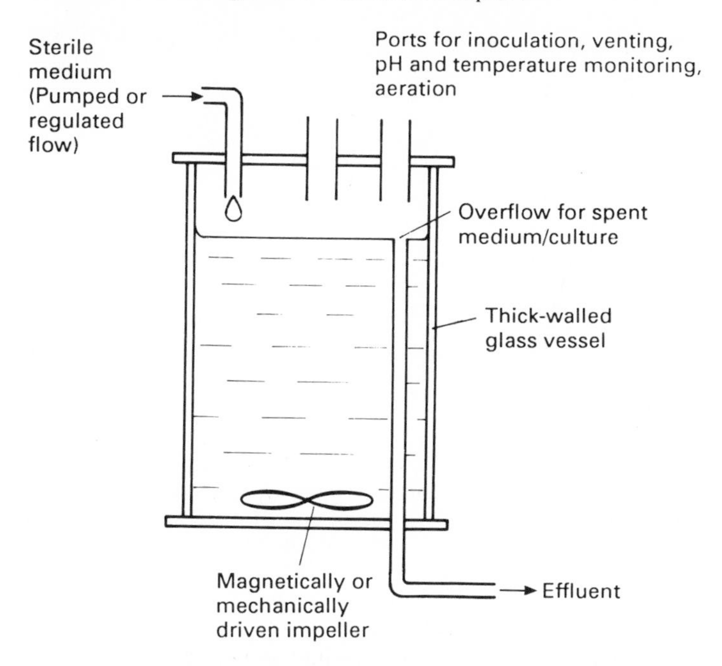

Fig. 36 A simplified diagram of a continuous culture system. Essential features are the continuous supply of new medium and its removal by overflow (an alternative is to pump out spent medium, using a tube which dips into the culture from above to remove medium when the level rises high enough to make contact with the tube). Also essential is a stirring mechanism to ensure homogeneity of conditions throughout the culture.

flow of fresh medium; a threshold arrangement such as an overflow for the removal of spent (or rather mature) culture; and monitoring equipment to check, for example, pH or temperature.

Continuous culture systems offer remarkable opportunities for the study of growth processes, as well as for continuous production of biomass of constant composition. If medium is fed in, and allowed to overflow, at a constant rate which is slow enough to allow microbial growth to continue after inoculation (i.e. not so fast that cells are washed out more rapidly than they can divide), a steady state is always reached in which organisms divide at a constant rate which just maintains their concentration at a constant level. That level will depend on the concentration of limiting nutrient in the medium as well as on the dilution rate employed, as shown in Fig. 37. The growth rate of the organism in steady state actually equals the dilution rate, provided dilution rate is equal to, or less than, the maximum possible growth rate in the medium used. At one extreme, when dilution rate equals maximum growth rate, we can imagine cells, at whatever concentration they were inoculated, doubling in number in the same time that it takes for the entire volume of medium to be replaced. In that time then, a number of cells equivalent to the starting population is removed, and the same number of new cells is produced by division. In practice, such a system would be very difficult to maintain and if the dilution rate was set slightly too high, cell numbers would fall steadily. Realistically, dilution rate must be slower than maximal growth rate for a steady state to be reached. Cells will grow initially at their maximal rate until their

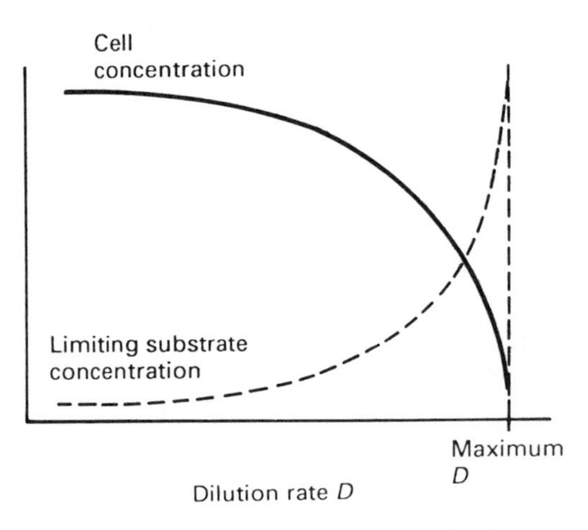

Fig. 37 The effect of varying the dilution rate on cell concentration and limiting nutrient concentration in a continuous culture. At the maximum sustainable dilution rate, cell concentration is minimal and nutrient concentration is barely affected.

numbers are increased sufficiently to reduce nutrient levels to the point where one nutrient becomes limiting for growth rate. When this limited growth rate falls to the same value as the dilution rate, the steady state is reached.

Continuous culture systems can be used in two ways – either as turbidostats or as chemostats. In a chemostat, the steady state is self-regulated as described above and controlled by setting the dilution rate. In a turbidostat, an optical cell and feedback mechanism to a pump supplying fresh medium maintains cell concentration at a predetermined level. For stable operation of a chemostat, dilution rates and nutrient levels tend to be low and cell concentrations high, while for turbidostats, high dilution rates and nutrient levels, with low cell concentrations, are more stable.

Continuous culture systems can be used as very powerful tools to investigate the interdependent effects of nutrient levels, temperature, growth rate and other factors on the growth process and its energetics.

The cell-division cycle: growth of individual cells

So far in this chapter we have considered the growth of cell populations – by far the easiest aspect of microbial growth to study. It is useful to consider however the growth of individual cells, and within limits there are methods for study of events during cell growth. There is little we can do by examining individual cells microscopically and watching them grow – we have no way of monitoring the molecular events taking place in a single cell during division. It may, however, be possible to monitor, for example, incorporation or radiolabelled protein or nucleic acid precursors for a brief period during growth as a function of cell size by microscopic examination of such cells exposed to a photographic emulsion so that incorporation of label into individual cells can be estimated. We can, however, usefully study such events in populations, provided all the cells of the population are at the same point in their division cycles: the cells are either selected for a particular point in the cycle or are grown in synchrony.

SELECTION OR SYNCHRONIZATION OF CELL POPULATIONS

There are several experimental approaches to this practical problem. The cell-division cycle affects several cellular parameters – for example, a cell about to divide will clearly be larger than one which has just done so, and therefore if we could separate all the largest cells from a culture they might all be about to divide together, i.e. in synchrony. To some extent it is possible to achieve this, for example by using differential filtration or centrifugal sedimentation in a density gradient. For such an experiment it is important that exponentially growing cultures are used, since in a stationary-phase culture there may be substantial variation in cell size in addition to that due to the position of the cell in the division cycle: cell age may also be highly diverse, and size may be affected by it. Provided a separation based largely on cell-division cycle parameters is achieved, the separated cells may either be analysed directly (e.g. for DNA content) or cultured to provide an initially synchronized culture in which all cells divide at approximately the same time. Synchrony unfortunately does not last long, however, as we shall see below.

An alternative approach to synchronization of growing cultures is to alter growth conditions in a pulsed manner, timed to encourage all cells to undergo cell division at the same time. For example, an essential metabolite for DNA synthesis might be added in limiting amounts at intervals to allow simultaneous chromosome replication in all cells, or cycles of raised temperature or the provision of light to a photosynthetic organism might be used to induce synchrony in the culture. Such methods are of limited value because cells may grow atypically in cyclically changing environments.

The most useful approach to the production of synchronized cultures has been the use of membrane filters to support bacterial cells which adhere to the filter, and are then allowed to grow and divide in a suitable medium at the optimum temperature flowing in a reverse direction through the filter (Fig. 38). Only newly divided cells will be released from the filter and appear in the effluent, and if these are collected over a short period relative to the length of the growth cycle, they can be used for experimental studies of the cycle.

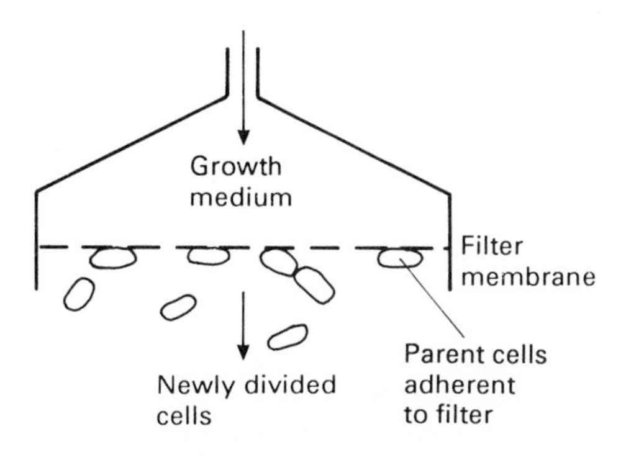

Fig. 38 The principle of the inverted filter method for isolation of newly divided cells. The bacteria are adsorbed onto the filter membrane and then allowed to divide in a stream of nutrient medium passed through the filter. After an initial period when loosely adherent mature cells may escape, the only cells which appear in the effluent are newly detached from adherent cells immediately after division.

EVENTS IN THE CELL-DIVISION CYCLE

Cell-division cycles in eukaryotic cells have been well studied and are familiar to most biologists, involving as they do the events of mitosis and the intervening periods of chromosome replication. In prokaryotes some features are similar, although of course the whole apparatus of chromosome division, nuclear dispersion and chromosome segregation is absent.

Basic features in prokaryotes are the synthesis of new DNA, followed by a necessary pause, and then by the cell division event. The most striking feature of the cycle is its flexibility: depending on the growth conditions, cells may divide extremely rapidly, or very much more slowly. They are able to develop extreme rapidity of division, despite the need for a pause after chromosome synthesis, by overlapping successive cycles of DNA replication under favourable conditions (Fig. 39). This is possible because the mechanism of chromosome replication is a linear process starting from a fixed origin on the chromosome, and once the replication 'fork' has moved away from the origin, a new round of replication can commence (Fig. 40).

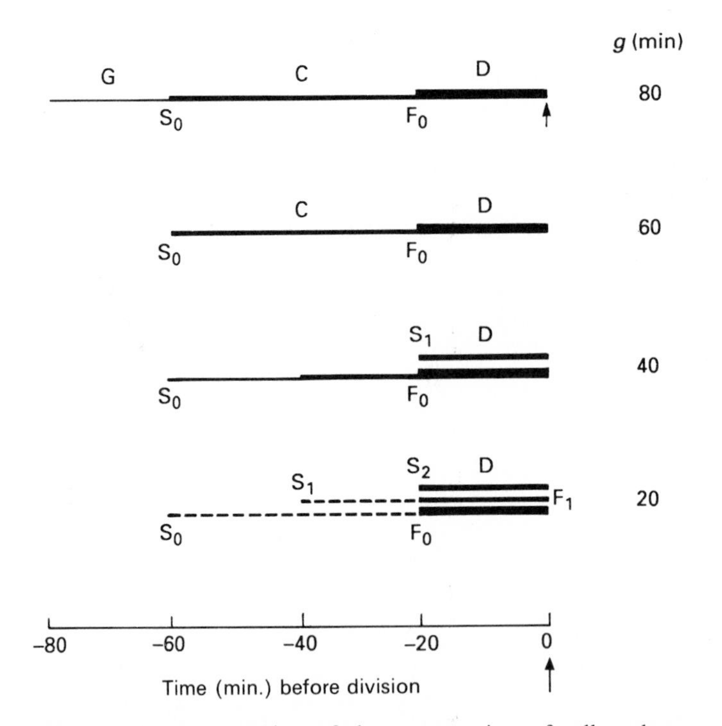

Fig. 39 A representation of the compression of cell cycle events and superimposition of successive cycles as the division time of cells is reduced. Thin line: cell growth period G, dispensable in fast-growing cultures. Intermediate line: chromosome division period C, shown dashed when commenced in previous cycle(s). Thick line: division of cell D. S: start of chromosome division, designated S_0 for the division relating to the 'current' D event, S_1 for the next, and S_2 for the one after that. F: finish of chromosome division. The arrow (time 0) indicates separation of the two daughter cells in the current division. It can be seen that at the fastest division or generation (g) time of 20 min, a chromosome division finishes (F_1); this will trigger the next D period, and thus D periods follow on one after another without pause.

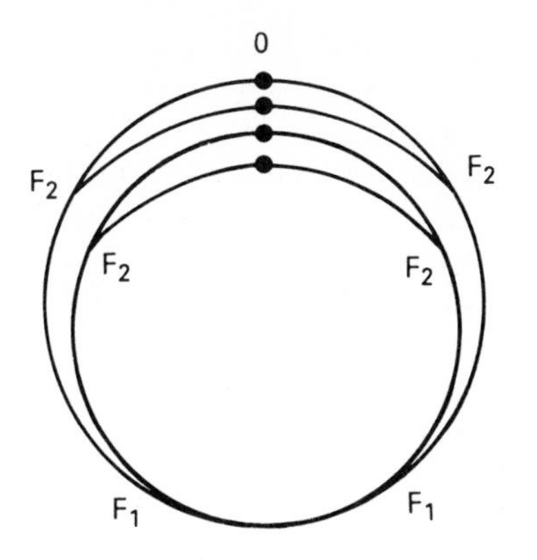

Fig. 40 Superimposed rounds of chromosome replication (bidirectional, commencing from the origin of replication O, and proceeding at replication forks F towards the diametrically opposite side of the circular chromosome). It can be seen that in fast growing cells with such superimposition, there are more copies of genetic material adjacent to the origin of replication than of genes at the opposite side of the chromosome.

Referring again to the typical *E. coli* cell, the main features of the cycle are a chromosome replication period (C) of about 40 min under favourable conditions, followed by a period (D) of about 20 min which is required for the run-up to division and its actual occurrence. In relatively slowly growing cells dividing about every hour, there would be an orderly sequence of chromosome replication followed by a pause, and then cell division (Fig. 39). During slower growth, there would be an additional pause after division but before chromosome replication, during which cell growth would take place. It is believed that the trigger for chromosome replication to be initiated is the attainment of a certain cell mass (or more precisely, the reduction of chromosome origin concentration per volume or mass of cytoplasm below a certain level), and that once this initiation occurs the process of chromosome replication takes place at a constant rate (assuming a constant temperature) – in other words the length of the C period is fixed. If rapid division is to occur, not only must the cells be large enough already at the time of division for a new cycle of chromosome replication to begin immediately, but in order to reduce the cycle time to less than the sum of (C + D), new rounds of C must commence *before* cell division occurs. By overlapping cycles of chromosome replication to the point where there are always at least two cycles under way (i.e. as soon as one finishes another commences – Fig. 39), the cell-division time can be reduced to the length of the D period. Effectively, this represents the maximum growth rate of the organism, since unlike chromosome replication the cell-separation events are not spatially organized in such a way that a new cycle can begin before the old one finishes.

GROWTH RATE AND CELLULAR COMPOSITION

There are important practical consequences of these cell-division cycle phenomena when we come to consider the growth of cells for experimental study. The most important is that the rate at which cells have been growing has a profound effect on their cellular composition and physiological activity. Fast growing cells are generally larger than slow growing cells. This is related to the overlap of chromosome replication cycles. Assuming that cell growth rate is governed only by nutrient availability, temperature, etc. then when newly divided cells are introduced to a more favourable environment, the time of their next division is already programmed by existing C and D cycles, and if their growth rate increases they will be bigger by the time they divide. Also, as cell size increases, the concentration of chromosome replication origins falls, and hence the initiation of new rounds of replication begins *before* division occurs. Although cell size is larger, and the total amount of DNA per cell is larger, the average concentration of DNA in the faster growing cell is lower: this is because while the concentration of origins of replication remains constant, the average size of the chromosomes in the cell falls – some of them are incomplete at any given time. We have to consider averages of cellular composition of course, because cells at different stages of the division cycle differ markedly in any case. The other major difference in composition of faster growing cells is RNA concentration, which is greater, and most of this increase is due to ribosomal RNA, accompanying the increased concentration of ribosomes. Another difference, consequent on greater cell size, is that since the ratio of surface area to volume decreases as cell size increases (see the Introduction), the concentration (as a fraction of cell mass or volume) of cell wall and surface components falls slightly.

In order to produce cells for study which are uniform in composition and size, we therefore have to ensure that they are all growing at the same rate; and if we wish to study cells in a particular state (e.g. the state in which they are found in a relevant natural environment), we may have to manipulate growth conditions to mimic the natural situation. One method of growth which does not favour cellular uniformity is growth on agar. Cells in the centre of colonies on agar will have their growth rate limited relative to that of cells at the colony edge, since nutrient concentrations will be lowered due to depletion by the growth of the colony. The most uniform conditions will be achieved in shaken or stirred cultures in liquid. The most reproducibly uniform cells will probably be obtained by taking exponentially growing cultures, and these are often favoured for critical experimental procedures. In the absence of deliberate manipulations to obtain such cells, we will often in the laboratory be studying cells in stationary phase – the product of either agar cultures or liquid cultures of relatively fast-growing organisms incubated overnight. The cells may be quite heterogeneous in their composition and physiological state.

CHAPTER 8

Practicalities of growth

The tools available to the microbiologist for growth of experimental microorganisms are relatively few, but the way they are used is very important. The simple classical methods of streak cultures on agar and static broth cultures in small volumes are still fundamentally useful procedures, for reasons which may not always be obvious. Nevertheless, the use of fermenter vessels, whether simple laboratory tools or highly sophisticated industrial plant, is a very powerful means of exploitation of microbial growth.

Solid or liquid media?

While solid media are far from ideal for production of microbial biomass, both for the theoretical reasons outlined in Chapter 9 and because of the sheer inconvenience of producing, inoculating and harvesting from large quantities of agar media, the agar plate has several powerful advantages for many aspects of the routine culture of microorganisms. We soon discover just how useful it is when working with an organism that is difficult or impossible to culture on agar, as for example with many of the spirochaetes!

CULTURE PURITY

Perhaps the most important use of the agar plate is to establish (at initial isolation), and subsequently monitor, the purity (freedom from contaminating organisms) of experimental cultures. This is dependent upon the production of individual, well-separated and reasonably numerous colonies (for choice of picking subsequently) by the microorganisms. Several methods of streaking cultures in order to achieve this aim have been described, some said to be more effective in circumstances where there are very large numbers, or a great diversity, of organisms. One method which works well in most situations, however, is shown in Fig. 41. It is reasonably beginner-proof and usually successful if two points are emphasized: (1) as much of the available area as possible should be utilized at each stage of spreading, by 'scrubbing' the agar surface rapidly but lightly with the loop rather than making a few separate streaks, except at the final step; and (2) by flaming the loop (and cooling it on the agar at the edge of the plate) between every sector, not just the first two. Beginners often make the mistake of

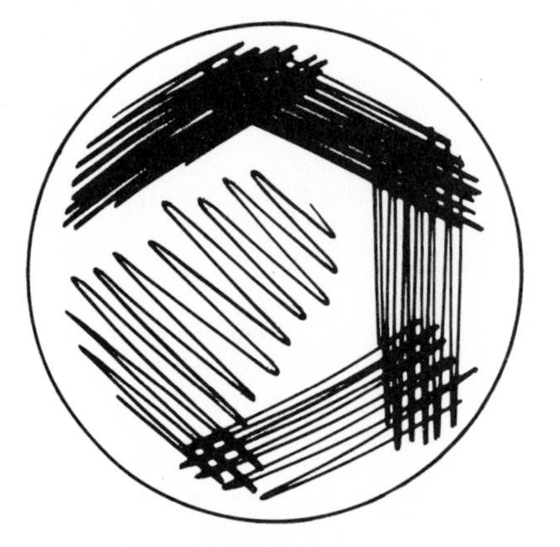

Fig. 41 Technique for spreading organisms with a loop to achieve numerous isolated colonies in at least one sector. However numerous or sparse the organisms, at least one sector will contain predominantly isolated colonies if the loop is flamed between sectors. Little of the space on the plate is untouched by the loop.

using too little of the area, with just a few streaks, so that perhaps only one streak will be inoculated at the optimum density and relatively few isolated colonies will be obtained. Well-streaked and badly streaked plates are shown in Fig. 42.

Whether isolates are to be picked from among a diversity of colonies after streaking (for example, a sample of richly populated sewage or mud), or whether individual colonies are to be picked from a stock culture (which might have become contaminated during storage) for further experimentation, careful examination of colonies is essential. Not only will different species appear either obviously or subtly different from one another, but surprisingly often subtle differences in colony type may reveal heterogeneity in experimental cultures of supposedly pure isolates. Textbooks often emphasize the variety of shapes and contours which microbial colonies may exhibit, and give them obscure and rather forgettable names. In practice these are rarely used, and because the exact character of the colonies formed by a species or variant may depend substantially on variable experimental conditions, the 'absolute' definition of colony character is not often crucial. What is important is the differentiation of one species or type from another, and this may also depend on size or subtle aspects of colour or other optical properties. Thorough examination by microscope – either a dissecting microscope or the low-power objective of an ordinary microscope, with a variety of lighting conditions – may reveal heterogeneity easily missed with the naked eye, and it is remarkable how colony morphology can be affected by a wide range of genetic changes in microbial cultures (see below).

The value of solid media in avoiding catastrophic contamination of cultures is obvious if we consider a contaminant organism, able to grow twice as fast as an organism being cultured experimentally. If one contaminant is included with an inoculum of 1000 organisms of the experimental culture, then after, for example, 12 doublings of the latter to yield about 10^7 cells, the contaminant will have doubled 24 times to about 10^8 cells, so it will now outnumber the organism being studied by approximately tenfold. In the analogous situation on solid medium, we would merely have one contaminant colony (admittedly rather a large one – although of course it

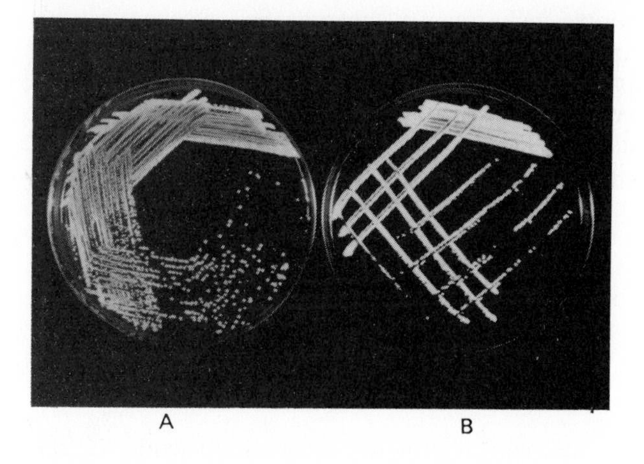

Fig. 42 Plates streaked effectively (A) and inefficiently (B). On plate B there are few isolated colonies and much wasted space on the agar.

may not be 10 times larger by the time the other colonies become visible because there is a limit to the availability of nutrients to support its growth) among 1000 colonies of the original culture.

IDENTIFICATION OF VARIANTS OR MUTANTS

The enormous power of this aspect of the use of agar plates is not immediately obvious, but consider the problem of isolating a mutant which, for example, may occur at a frequency of 1 per 1000 cells in a culture. To find such a cell in liquid cultures, assuming there was no selective or enrichment method available, we would have to perform limiting dilutions (see Chapter 9), inoculate perhaps 10 000 aliquots of medium to ensure a reasonable probability that those which grew were truly clonal, and test each of 1000 growing cultures for the property being sought. To take as an example the isolation of a mutant of the *Pneumococcus* which did not produce a polysaccharide capsule (a crucial step in the discovery of DNA as the genetic material), such an experiment might require the microscopic examination of organisms from thousands of liquid cultures to find one which did not possess capsules. If we were lucky, there might in fact be a difference in growth character in liquid as well as on solid media, so we might be able to search quite rapidly by eye, perhaps for a culture which settled out quickly to the bottom of the culture tube while wild-type cultures all remained uniformly turbid. If, however, we used agar media, capsule-negative mutants would appear rougher in texture, and perhaps smaller, than the wild-type, encapsulated parents, and we would be able to spot such colonies in minutes among perhaps several thousand spread over half a dozen agar plates. Similar considerations apply in the use of agar plates for viable counts (see Chapter 9).

An additional refinement of solid media which has been widely exploited for identification of variants or mutants is to make colony differences much more obvious, or to induce differences between colonies which are otherwise indistinguishable, by the modification of media so that clear differences, usually in colour, exist between wild type and mutant colonies. For example, inclusion of a 'chromogenic' enzyme substrate in the medium will lead to colour development around a colony of organisms which are capable of processing the substrate to release a coloured product. A well-known example is the beta-galactosidase substrate X-gal, which in the presence of the potent inducer of beta-galactosidase activity, IPTG, will produce blue colonies of cells which have the enzyme. A similar principle is used to identify lactose-fermenting colonies in the classical isolation medium for coliform organisms, MacConkey's agar. Lactose is present in abundance in this agar and when fermented, leads to production of organic acids which lower the pH of the medium locally around colonies. These therefore turn pink due to the presence of the indicator neutral red, more or less colourless at the neutral pH of the unused medium, but bright pink at a lower pH.

SELECTION OF SPECIES OR MUTANTS

Although both solid and liquid media can in principle be used efficiently for the selective growth of particular species or known mutants, solid medium has advantages in some circumstances and is often used for selection or enrichment. The former may be defined as growth in the presence of growth inhibitors or agents which are lethal for organisms which do not have the property being sought, whereas the latter is growth in conditions which selectively permit the rapid growth of particular organisms, perhaps by provision of a favoured nutrient which is likely to be used only by the particular organism sought. Solid media are very often used in molecular biology for the selection of antibiotic-resistant organisms, whether as mutants or to confirm or maintain the presence of a known antibiotic resistance gene, by incorporation of the relevant antibiotic in the agar medium before plates are prepared. Again, there is the advantage that individual antibiotic-resistant clones (colonies) can be picked immediately and individually from antibiotic-containing plates for further characterization.

There are additional – and often quite complex – applications of the two-dimensional format of multiple colonies on agar plates, often allowing very rapid and economical means of analysis of multiple microbial clones. A classical example is the technique of *replica plating*, in which we may, rather than wishing to select antibiotic-resistant clones of bacteria, need to identify susceptible clones among a majority of resistant organisms. This can be done by producing agar cultures of suitable colony density, and then inoculating replica plates from them by means of a suitable sterile solid matrix which will pick up a little of the growth from each colony, and re-inoculate it onto additional plates which may contain the antibiotic of interest. Velvet stretched over a wooden or plastic former is a traditional tool. Organisms from parent colonies of susceptible organisms will not lead to growth of replica colonies on the replica plate (Fig. 43). Developed from this, biochemical or immunochemical properties or presence of specific nucleic acid sequences may also be monitored on similar two-dimensional replicas, prepared either by overlaying a disc of nitrocellulose paper or other support film onto the colonized parent plate to pick up material from each colony, or by growing colonies directly on nitrocellulose and taking a replica onto fresh medium, for future use, before processing the disc. A visual analogue of the colony pattern is prepared by means of colour reactions for enzyme activity, or immuno-labelling methods for detection of antigens, or nucleic acid hybridization tests for specific sequences, all yielding spots of distinct colour or intensity depending on the presence or absence of signal from the colony replica. Again, the two-dimensional analogue

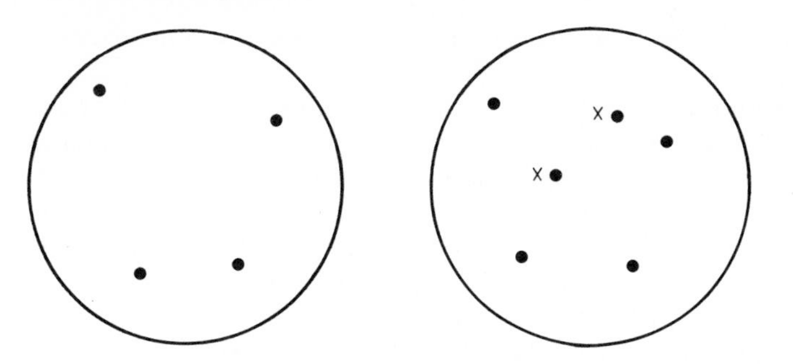

Fig. 43 The principle of replica plating. Colonies from the plate on the left are reproduced in an analogous pattern on the right; any which are unable to grow (e.g. under a selective pressure in plate B), can be identified by their absence and the organisms recovered from the original plate at X.

provides an immensely powerful tool for rapid screening of large numbers of clones.

One more advantage of solid media for handling large numbers of clones is their suitability for spatially separating large numbers of cultures without the need for separate containers for each. A common system for temporary holding of clones while, for example, they are tested in some assay system, is to subculture colonies in an organized grid pattern, perhaps as many as 50 or so on one plate. Again, replicas may be prepared with less risk of closely adjacent colonies running together, which is a limitation in direct replica plating.

COUNTING

Solid media confer a surprising degree of advantage for the counting of viable organisms, since one agar plate, with space for well over 1000 colonies to develop without touching or merging (provided they are not too large or flowing), is therefore the equivalent of perhaps 1000 culture tubes or other replicates in a quantal assay such as a limiting dilution count (see Chapter 9). This argument holds even if we only have 30 colonies on a plate – we can argue that these correspond with 30 out of 1000 tubes showing growth in a quantal assay (which of course would never be used with such large numbers of replicates, for practical reasons). The plate of solid medium is thus uniquely powerful for counting purposes, and equivalent colony-based methods have been developed, for example in virology where a plaque in a cell monolayer is the equivalent of a colony on agar.

GROWTH OF FASTIDIOUS ORGANISMS

Certain organisms which are quite simple to grow in solid media may be difficult to grow, or to grow in the right way to produce important products, in liquid media. An example is the whooping cough organism, *Bordetella pertussis*, which requires very specialized liquid media for the production of organisms with the full complement of antigens necessary for whooping cough vaccine and was grown on solid media for vaccine production for many years. Similarly, production of surface antigens such as fimbriae of Gram-negative organisms may be dependent on growth on solid media.

Advantages of liquid media

Despite the numerous applications of solid media outlined above, there are of course important applications of liquid media as well. Their use, with proper handling and adjustment of culture conditions, is essential for production of organisms of uniform properties for critical experimental procedures in biochemistry or molecular biology. As outlined above, logarithmic phase cultures are often used, for uniformity as well as for their optimal properties in other respects. It is not easy however to produce a true log-phase culture to order. If a flask containing say 50 ml of rich medium such as L broth is inoculated with a small (barely visible) particle of colonial growth from an agar plate culture of *E. coli*, and shaken at 37°C in air, by the time the colony sample is suspended evenly after a few minutes shaking, the 10^7 or so organisms present will be diluted down to $10^5 - 10^6\,ml^{-1}$. Even at the lower limit, there will only have to be about 4 logs increase, or 12 doublings, to bring the number up to about the maximum supportable by such a medium. This will take only about 4 h. The usual practice is to grow such cultures overnight, then to subculture into fresh medium to obtain a known optical absorbance, and monitor it closely to predict the time of the late logarithmic phase of growth. If started at an absorbance of 0.1, it can be allowed to rise to about 0.5 before there is likely to be any dimunution of growth rate as the end of the log phase approaches. This is between two and three doublings, which if growth began immediately would take only about 40 min. In practice there will be a lag phase of perhaps 30 min, but frequent monitoring of absorbance every few minutes will be necessary to 'catch' the culture at the right moment, and a coffee-break taken at the critical time may ruin the experiment!

Use of fermenters

To obtain the highest possible yields of organisms in liquid batch culture, fermenters are often used since the control of culture conditions can be very sophisticated. The basic needs are shown in Fig. 44: a stirred or agitated vessel, which may be achieved by impeller, magnetic stirrer, sparger or airlift; temperature control, including both heating and cooling capacity, the latter so that cultures will not overheat; pH control, involving both a pH electrode to monitor and an injection system, normally for alkali to be added to reduce metabolic acidity; and ports for inoculation and sampling. The operation of fermenters is a skilled science in itself. Careful control of stirring speeds and aeration may be necessary to prevent foam build-up and destruction of sensitive proteins or other cell products by denaturation at surfaces by surface tension. Chart recorders may be used to record the culture parameters at different times during growth. Because of the increased control over conditions which these facilities provide, it is possible to obtain very high yields of cultures under fermenter conditions, with concomitant advantages for downstream processing due to the fact that culture products are highly concentrated before processing begins.

Growth of anaerobic bacteria

There are two categories of anaerobic organism, and two kinds of technical approach to handling them. The first is of oxygen-tolerant organisms, able to survive in the

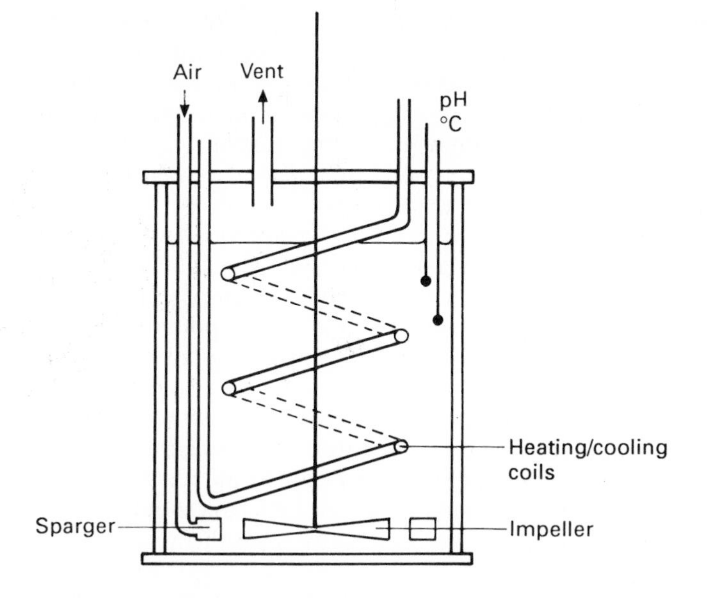

Fig. 44 The basic facilities required in a fermenter. The culture must be stirred for homogeneity of conditions, and the temperature, state of aeration, pH, etc. monitored and controlled if necessary by feedback mechanisms.

presence of normal atmosphere at least for long enough to be handled during subculturing and other growth manipulations, without being killed by the presence of oxygen. Spore-formers such as *Clostridium* may be in this class, in part because the spore is resistant to the effects of oxygen, but vegetative cells of some species also seem to be more tolerant than others. These tolerant species can be handled on the open bench without special precautions and merely incubated in appropriate media and gaseous environments; attention to technique can improve the success rate. The second category is of oxygen-sensitive organisms, sometimes extremely so, and these must be isolated and handled with extreme precautions to prevent exposure to air.

OXYGEN-TOLERANT ANAEROBES

The basic need is for anaerobic incubation facilities, and these have been described and illustrated briefly in Chapter 2. For practical purposes, incubation is best done in sealed jars with gas-generating packs and catalyst to purge residual oxygen. A note about the catalyst: manufacturers supply these in pellet form, for use in metal gauze packs suspended from the lid of the jar, and instructions for their use should be followed carefully since they will become exhausted quite quickly. One precaution may be needed – if jars are overloaded so that petri dishes come into contact with the catalyst gauze, sufficient heat may be generated during the reaction of hydrogen and residual oxygen to melt the plastic! Otherwise the systems are reliable and efficient, although a little expensive on consumables if used without careful planning to minimize the number of packs used – they currently cost about 50 pence each.

Media are usually complex, since many anaerobes are also fastidious, and serum and other supplements are commonly added. It is advisable, if not essential, to keep liquid media free from contact with, or exposure to, oxygen as much as possible. Autoclaving drives off dissolved atmospheric gasses, and if autoclaved with small air spaces and loose caps which are tightened immediately on removal from the autoclave, media will remain quite oxygen-free for storage until use. They should also be kept in the dark, to prevent light-catalysed oxidative processes from generating potentially inhibitory products. Reducing agents such as thioglycollate or cysteine are also used to lower redox potentials of media, but these also may mediate the generation of toxic free radicals if they do come into contact with oxygen, and they should be used with care, in conjunction with exclusion of air as described above. The redox potential actually represents the summed effects of all redox couples in the system, so there is potential for either low redox potentials with some oxygen present, or high redox potentials in the absence of oxygen. Oxygen level is far more important than redox potential *per se* in determining the survival and growth of strict anaerobes.

If supplements are added to media after autoclaving, they are likely to be labile, filter-sterilized solutions and will not therefore be oxygen-free. Therefore, the complete liquid medium with added supplements should be pre-reduced, or purged of oxygen, by pre-incubating in an anaerobic jar overnight before use. Solid media should also be pre-incubated – obviously there is no way to keep air away from them without an oxygen-free working environment.

Oxygen-tolerant species may be handled without special precautions to exclude oxygen during streaking, diluting, etc. Oxygen-sensitive organisms can also to some extent be handled successfully if manipulated rapidly and correctly, especially in liquid culture. If pre-reduced media are inoculated with quite large volumes of inoculum (e.g. 10 per cent by volume) as a suspension, so that containers are full, some anaerobes can actually be grown without any special gas phase if the cultures are sealed. The large inoculum helps to purge any residual oxygen through its metabolic activity (it is worth remembering that it is actually microbial activity which removes oxygen from polluted rivers and from our own large intestines to generate environments where strict, oxygen-intolerant anaerobes can thrive!).

Strict exclusion of air

HUNGATE TECHNIQUE

This technique utilizes commercially obtained gases such as nitrogen or argon, freed of all traces of oxygen by passing over heated copper turnings, to replace the atmosphere in all culture vessels by directing a stream of gas into all vessels when they are opened for access (Fig. 45). Stoppers of butyl rubber are commonly used to maintain the gaseous environment within the culture vessel. For liquid media this is straightforward: liquids can be transferred to inoculate cultures in gas-flushed pipettes or syringes; for solid media, the equivalent of the petri dish is a roll tube, in which agar medium is introduced into a tube which is rolled mechanically during setting so that a thin layer coats the inside of the tube. Inocula can again be introduced as small volumes of liquid, but streaked by allowing a loop to draw a little of the inoculum up in a shallow spiral round the tube towards the neck by holding the loop in position while the tube revolves, again on the mechanical roller apparatus. Isolated colonies should grow near the top of the tube.

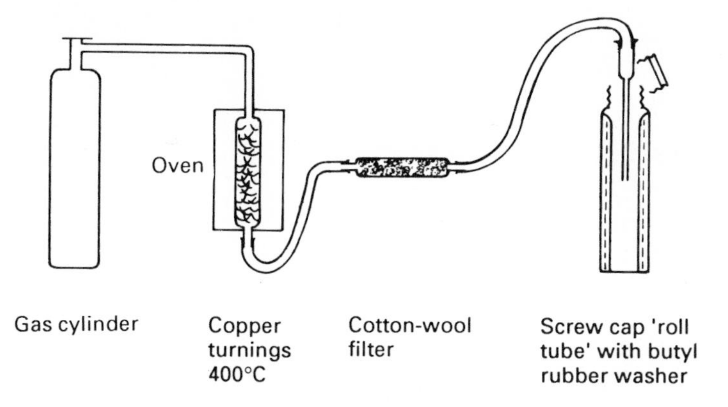

Fig. 45 Diagrammatic illustration of the Hungate technique for maintenance and culture of oxygen-intolerant anaerobes. Organisms are kept in sealed vessels with an oxygen-free atmosphere, and all manipulations are done under a stream of oxygen-free gas from which all oxygen is purged by passage over, for example heated copper turnings. The cotton-wool filter is to ensure sterility. The roll tube is analogous to a petri dish – an agar surface is established on the inside of the tube by setting molten agar while the tube revolves on rollers, and it can be inoculated under the oxygen-free atmosphere with a loop.

To indicate whether redox potentials are low, which normally would not be compatible with the presence of oxygen, the indicator resazurin is often used. This is colourless when fully reduced, and turns pink if the redox potential rises significantly, usually due to the presence of oxygen.

ANAEROBIC CABINET

Although the equipment needed is complex and expensive, this is probably the most convenient technique for the isolation of extremely oxygen-sensitive organisms. The essential features are shown in Fig. 14 (see p. 21). An atmosphere of nitrogen:carbon dioxide:hydrogen (80:10:10) is maintained in the working space, to which access is gained by means of an air-lock which can be evacuated and the air replaced with the gas mixture. Recirculation of the atmosphere through a 'scrubber' ensures removal of residual oxygen. Access to the working space is gained via arm-length rubber gloves sealed into ports at the front. Media (solid and liquid, ready to use and equilibrated with the anaerobic atmosphere) are stored in the working area. An integral incubator may be included, or alternatively cultures may be placed in anaerobic jars for removal. Bunsen burners cannot of course be used: a small electric heater is an alternative, or disposable loops may be used for inoculations, culture transfers, etc. Aseptic technique must be performed carefully since flaming cannot be part of it.

If strict and oxygen-sensitive anaerobes are to be isolated from environmental or clinical sources, it is important that attention is paid to exclusion of air right from the start. Pre-reduced transport media, or gas-filled containers and anaerobic jars, should be used otherwise there may be substantial losses of the most sensitive organisms before they even reach the lab.

CHAPTER 9

Microbial quantitation

Quantitation of microbial populations is a crucial aspect of experimentation, for the basic reason that populations rather than individuals usually make up the experimental material used by microbiologists. Methods of quantitation vary widely both in the methods used and in the principles behind them, and each has significant advantages and disadvantages which are listed in Table 8, and will be described in detail below.

Several methods of quantitation involve the preparation of serial dilutions, and a good grasp of the principles and practice of this procedure is essential. Wrongly or badly used, it can produce some startling results. A paper submitted to a scientific journal recently claimed a viable count in a liquid culture of $10^{15}\,ml^{-1}$ – not merely impossible to support with the nutrients present, but a physical impossibility to fit into the space available!

Serial dilution

Serial dilution is the repeated dilution of a liquid suspension or solution, in a series of separate steps, so that a very large dilution overall can be achieved in a manageable volume by serial transfer of small volumes sequentially through a series of tubes or bottles of diluent (Fig. 46). Because the series is based on a sequential reduction of the concentration by a constant factor, rather than on a series of additions of equal proportions of diluent, the dilution series is geometric rather than arithmetic, and considerable dilution factors are obtained rapidly – in a logarithmic fashion in fact. Figure 46 shows twofold, fivefold and tenfold dilution series, as well as an arithmetic series covering a tenfold range for comparison.

If we consider the tenfold series, it can be seen that in only six steps, a dilution factor of 10^{-6} or 1/1 000 000 is obtained. Obviously to achieve this in one step, we should need to add, for example, 1 ml of starting suspension to 1000 litres, or 1 tonne, of diluent, whereas by the tenfold serial dilution method we can achieve this with the use of 60 ml of diluent starting from 1 ml of suspension. Note the use of approximations here. Actually, it is important to grasp firmly the exact way a dilution factor is obtained: a tenfold dilution is obtained by adding one volume of suspension to *nine* volumes of diluent, *not* to *ten* volumes, so the volumes needed in the above example would be 999 999 ml for a single-step dilution, and $6 \times 9 = 54$ ml for the six steps in a tenfold

Table 8 Methods for quantitation of bacterial suspensions.

Method	*Sensitivity*	*Advantages*	*Disadvantages*
Visual (chamber)	$>10^7$ ml^{-1}	Immediate	Laborious for replicates; accuracy poor
Optical (absorbance)	$>5\times10^7$ ml^{-1}	Immediate; accurate	Total particles counted, dead or alive; sensitivity poor
Dry weight	$>10^{10}$ total	Absolute value	Slow, insensitive
Chemical (e.g. protein nitrogen)	10^6 approximately	Absolute value; accurate; sensitive	Laborious; affected by physiological state of organisms
Viable count (plate)	10^2 ml^{-1}	Sensitive, counts live cells only	Delay for growth; may not give separate colonies for, for example, clumped cells
Viable count (limiting dilution)	10^2 ml^{-1}	Useful, for example, for organisms unable to grow on agar	Delay for growth; highly inaccurate; media and equipment needs excessive

series. The reason why the volume needed is only 9 ml for each step in a tenfold series is that the dilution factor, i.e. the concentration of the suspension after dilution compared with its concentration before dilution, depends on the initial volume compared with the final volume occupied by the same number of particles. The latter is made up from the volume of diluent *plus* the volume transferred: 9 ml of diluent *plus* 1 ml transferred, i.e. 10 ml. In the usual jargon, we often speak of a tenfold dilution as 'one in ten' or 1:10, which indeed is the mathematical value of the dilution step – 1/10, or 0.1, or 10^{-1} ('1 log'), or the ratio of starting to finishing volume. Initially, this can be confusing, since in making the dilution we actually add 1 ml into 9 ml of diluent, which seems more like a 'one in nine' dilution.

Note the use of the power of 10 or logarithm of the dilution factor. Especially in the use of tenfold dilution series, but also with twofold series (in which case we should use $\log_2$), it is very convenient to express dilutions as the log value. Dilutions of 1/100000000 may not be uncommon: obviously this is clumsy to express in such a way, and it is easier and causes less confusion to use a $\log_{10}$ value, which would be -8 in this case. As emphasized in Chapter 1, the student should become used to thinking in terms of such log values when dealing with large numbers or high dilutions.

Depending on the practical application, the value of each dilution step in a series may vary. If high counts, i.e. high concentrations of organisms are present in a suspension, it is not practical to use anything other than a tenfold dilution series to obtain, for example, dilutions of 10^{-5}, 10^{-6} and 10^{-7} for counting purposes (see below). For lower concentrations, however, where the approximate count may be known, it may be practicable to use a fivefold or even a twofold series, and a more accurate determination with a smaller cumulative error may be achieved.

Cumulative errors are one of the major problems in performing a serial dilution. They are serious because their effect is multiplied at each step in the series. Thus an error of just 10 per cent in the volume transferred (i.e. 1.1 ml instead of 1.0 ml in the above example) will lead to a 40 per cent difference in the dilution obtained after only five dilution steps. An even more serious problem in making serial dilutions can be

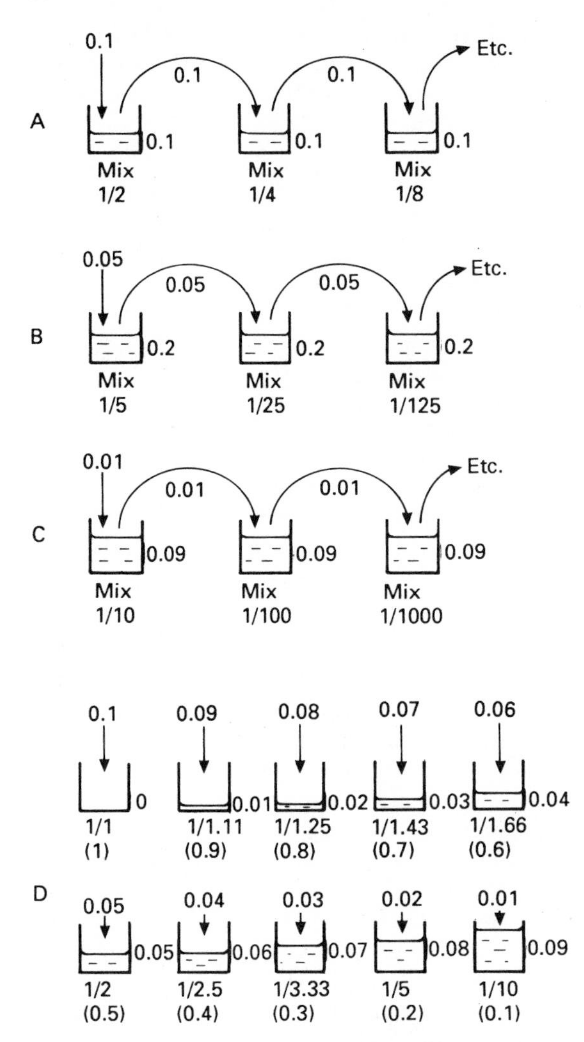

Fig. 46 Some examples of dilution series; A–C are geometric, D is arithmetic. Arbitrary volumes of diluent are shown next to the wells, and volumes transferred or added above the wells or tubes. The volumes could alternatively be any constant multiple or fraction of those shown. (A) twofold (each step 1 in 2); (B) fivefold; (C) tenfold; (D) a linear series from undiluted to 1 in 10. Final dilutions obtained are shown below the wells. In the linear series, the concentration of the diluted substance is shown also as a decimal fraction (bracketed).

caused by the use of a single pipette, or pipette tip, in performing the series of steps. Provided *all* the liquid carried over each time were thoroughly mixed at each step with the successive volumes of diluent, there would be no problem. However, if a small volume of concentrated suspension from an early dilution is deposited high up in the pipette at an early stage, not mixed with the diluent, and then mixed in several steps later, perhaps because it has run down the pipette, it may cause an enormous error even if the volume involved is tiny. Even 1 μl of starting suspension, introduced into 10 ml at the 10^{-5} dilution stage, will increase the concentration of the suspension at that dilution by *tenfold* above the value it should have. Such errors, although not necessarily so extreme as this example, are almost inevitable if a long dilution series is made without changing pipettes, and again they can be cumulative along the series.

USE OF MICROPIPETTES

Modern laboratories rely increasingly on micropipettes and microtitre dilution systems, and it is important to understand their operation. Most operate on a similar principle, illustrated in Fig. 47. Because surface tension is strong enough to withstand a significant part of the air pressure which expels liquid from the tip when the piston is depressed, provision is made for extra movement of the piston to a second stop, beyond

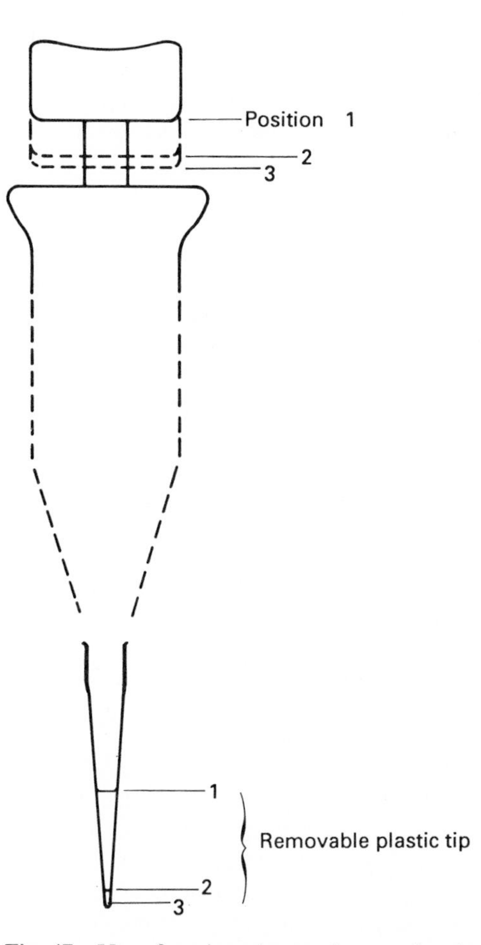

Fig. 47 Use of a micropipette. At rest, the plunger is held by a spring at position 1. There are two 'stops', reached when the plunger is depressed lightly to position 2, and more strongly to position 3. The measured volume is drawn into the tip by releasing the plunger from position 2 to 1. It may be expelled by depressing the plunger to position 2, and in addition, if necessary to expel a final drop held in the tip by surface tension, to position 3. For mixing contents of tubes or wells, the plunger can be depressed repeatedly to position 2 and released. Then all liquid should be expelled with the tip clear of the solution by depressing to position 3, before again drawing up the measured volume between positions 2 and 1.

the volume required, so that the last tiny volume of liquid can be expelled. To take up the fixed volume, only the correct volume from the first stop should be withdrawn. Thus in making serial dilutions, it is vital that the tip is withdrawn from the liquid after mixing (done by repeated depressions of the piston to the first stop only), all fluid expelled by use of the second stop, and then the final volume withdrawn carefully using the first stop only. Students sometimes do not understand the use of first and second stops and use both at every depression of the piston.

Methods of counting

Actual counts of real numbers of organisms, rather than measurements of parameters dependent on counts, have two particular advantages: (1) there are methods for counting both the number of viable organisms (not always all the organisms in a culture will be alive) and the total number – in different circumstances either may be desired; and (2) the count obtained is independent of other variables in cell suspensions.

PLATE COUNTS

Plate counts are essentially direct counts of numbers of colonies in, or on, agar plates, each colony being assumed to represent the progeny of one organism. A possible error which immediately arises is that if two or more organisms are stuck together, colonies will not necessarily arise from single organisms, and often plate counts are recorded in terms of 'colony forming units' or c.f.u. to acknowledge this. Table 8 shows that while the method has a good sensitivity, being able to quantify low concentrations of organisms, it is normally dependent on serial dilutions, with the attendant potential for error. The sensitivity can, incidentally, be increased dramatically by combining it with filtration so that, for example, water samples with very low counts can be tested by passage of a substantial volume through a membrane filter which is then laid on to an agar plate so that nutrients diffuse through the membrane and allow growth of colonies directly on its surface.

Following the preparation of dilutions, normally in a tenfold series, from a concentrated microbial suspension, plates may be inoculated in three main ways: *spread plate, drop plate* and *pour plate.*

Spread plate The most commonly used is the spread plate, in which a fixed volume of dilution, usually about 0.1 ml, is deposited on the agar surface and spread over it by simple glass spreader (Fig. 48) which is conventionally sterilized by dipping in 70 per cent ethanol and flaming to remove the residual solvent. Several technical problems are common. A large volume of the inoculum may be pushed to the edge of the plate, where it is inefficiently spread and initiates growth of large numbers of colonies, too close together to count. Contaminant colonies, which may be tiny, may be present on the plate before use and if not noticed will be spread round the plate so that their progeny may obscure the colonies being counted. More commonly, the plate may not be dry enough, and this is crucial because if the liquid of the inoculum is not all absorbed immediately, organisms may start to divide in the liquid phase and may ultimately produce many colonies from one cell. Fortunately, this phenomenon is easily detected

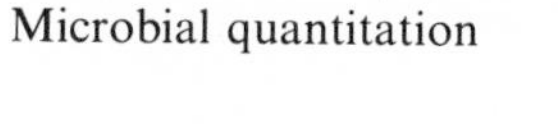

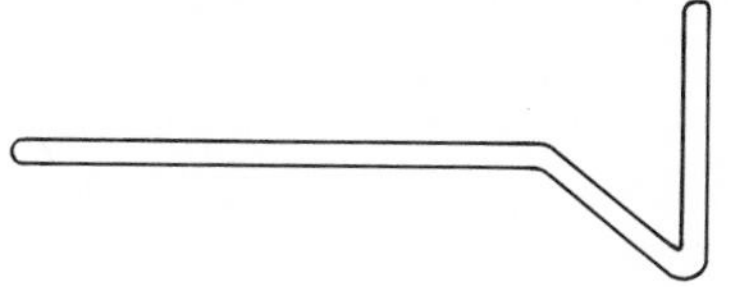

Fig. 48 A simple spreader made from glass rod bent in a flame. It can be sterilized effectively by dipping in 70 per cent ethanol and burning off excess. Care is needed not to set fire to the ethanol used for dipping, which should be positioned well clear of the flame. It burns with a near-invisible flame and is therefore especially hazardous if not noticed.

because the pattern of distribution and size of colonies is atypical: they will tend to be produced in elongated clusters and will often be unusually small (Fig. 49). It is also important that plates are properly prepared – uneven thickness or over-drying at one edge may lead to part of a plate inadequately supporting growth, especially with more fastidious organisms. An obvious question about the spread-plate technique is whether a significant number of organisms remain on the spreader; in fact the number which do so is surprisingly small, and can be neglected.

Drop plate Related to the spread-plate method is the drop plate, prepared by simply depositing small drops of inoculum suspension, distributed over the plate (Fig. 50). Formerly this was often done with a fixed-volume glass pipette delivering 0.1 ml, touched onto the agar at perhaps a dozen points while gently and slowly expelling the liquid; alternatively, a calibrated pasteur pipette could be used, and drops of liquid of known volume allowed to fall from the tip. Nowadays, a micropipette can be used to deliver repeated drops of known volume accurately. On well-dried plates, as much as 0.2 ml in one drop may be distributed around by repeated tilting of the plate. In all cases rapid soaking-in of the liquid is crucial and plates must be well dried. The method is very rapid since no spreading is involved, but colonies may be inconveniently close together for easy counting. On the other hand, there are no losses due to adherence to the spreader, and plate contaminants will not be spread.

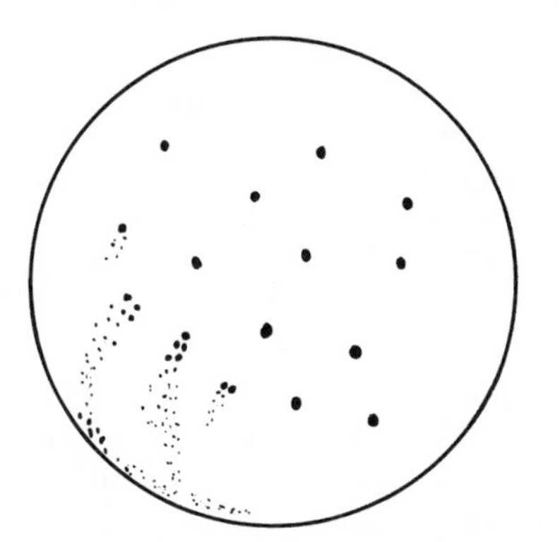

Fig. 49 The effect of inadequately dried agar plates on colony counts. Where growth has taken place before the liquid was all absorbed into the agar, small colonies are spread across parts of the plate in characteristic patterns.

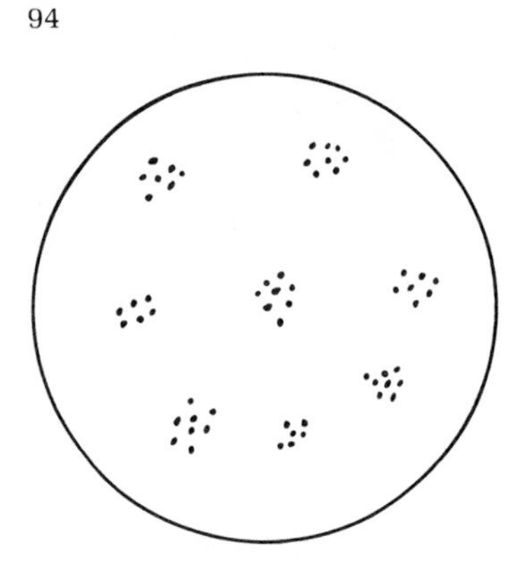

Fig. 50 The colony pattern generated by drop inoculation for counting.

Pour plate The third well-known method for plate counts is the pour plate. A known volume of inoculum suspension is mixed with about 20 ml of liquid agar medium held in a water bath at the lowest practical temperature to prevent it setting, usually about 45°C. After mixing rapidly, but gently to prevent bubble formation, the contents are poured immediately into a petri dish and allowed to set. The plate is then incubated, and colonies grow within the agar and can be seen easily, often as lens-shaped particles. The method avoids problems due to wet plates, and also may be preferable for organisms which are susceptible to lysis by surface tension – this may occur with some organisms if in a poor physiological state, for example if just recovered from storage frozen. Disadvantages are that transparent media must be used (blood agar, for example, could not be), and the procedure is somewhat slow and cumbersome. There is also a danger that exceptionally sensitive species may be adversely affected by the sudden temperature increase on mixing with medium. Aerobic organisms may grow slightly more slowly than on spread or drop plates, since gaseous diffusion through the agar may be a limiting factor.

Numerical aspects of plate counts For maximum reliability, plate counts should be arranged so that results are statistically valid as well as convenient to perform. In general, this means that plates which have fewer than about 30 colonies should not be relied upon, since the possibilities for error due to random fluctuation in number are substantial. For convenience of counting, the ideal is no more than about 300 per plate, since above this number the counting takes too long and colonies may be too close together for reliable counting. In order to be sure that at least one plate or ideally more than one replicate plate will have the ideal number of colonies for accurate counting, several dilutions will usually have to be plated – often three tenfold dilutions can be used to cover the expected range. Various methods are used to assist counting – from simple use of a hand-held push button tally counter, marking the outside of the plate with a felt-pen to show which colonies have been counted, through the use of electronic counters which work either through pressure activation of a switch as a pointer or pen is pressed to the outside of the plate over the colony, or by completion of an electrical circuit as a metal probe is touched to each colony, the circuit being completed by another probe inserted into the edge of the agar. Even more sophisticated are computerized methods for producing and analysing a digital image of the plate which can be counted automatically by image analysis.

Table 9 Results obtained from a 'typical' set of plates counted for a viable count determination.

Dilution	*Numbers of colonies (duplicates)*	*Mean*
10^{-4}	538, 502	520
10^{-5}	60, 70	65
10^{-6}	5, 9	7

CALCULATION OF VIABLE COUNT

The calculation is straightforward, but students often make a simple error: they forget to correct for the volume plated or the dilution factor. The correct formula, where the mean number of colonies is N for a volume plated of V ml at optimum counting dilution D, is

$$\text{Viable count (ml}^{-1}) = \frac{N}{V \times D}$$

A typical set of results is shown in Table 9, and one or two questions arise about their interpretation. Note that whereas there should be a tenfold difference between dilutions, the mean values for the countable plates only differ by a factor of eight. This kind of error is very common, and is usually due to the carry-over factor described above. Random variation in the rather small numbers of colonies at the higher dilution of 10^{-6} could be responsible for the larger difference between them, and these figures are best disregarded. Moreover, there is one additional dilution step compared with the values at the lower dilution of 10^{-5}, and the latter are likely to be more reliable for this reason also. The difference between values at 10^{-4} could be due to overcrowding of colonies. Therefore the count is best based purely on the values at 10^{-5}, and is:

$$\text{Viable count} = \frac{65}{0.1 \times 10^{-5}} \text{ml}^{-1}$$

$$= 6.5 \times 10^{6} \text{ ml}^{-1}$$

Note that in expressing values of large numbers in this way, it would be illogical to give the first number a value anything other than something between 1 and 10, since the advantage in the use of powers of ten, of avoiding confusion with, say, numbers of noughts or the position of the decimal point, would be lost. Thus the above count could equally be expressed as 0.65×10^{7}, or 650×10^{4}. Use of such forms should be avoided. Note also the units of the count (ml^{-1}) which should always be included (and might remind us to correct for the volume plated), and it is preferable to use the mathematically correct form, 'ml^{-1}', rather than 'per ml' or '/ml', to avoid confusion.

Sometimes there may be an unacceptably large difference between replicate plates or discrepancies between successive dilution steps in the performance of plate counts. It is instructive to consider the best way to remedy this. One method would be to increase the numbers of replicates, so that the mean values would have greater statistical reliability. However, this may not be the best solution: suppose the dilution process itself were the source of problems, in which case increasing the number of replicates would not help to give a more reliable value and discrepancies between dilution steps would remain. Instead, it may be better to duplicate the dilution series. Thus careful interpretation of the likely causes of erratic results may indicate the best remedy. For greatest accuracy, each step in the experimental procedure should be considered in turn, and the likely sources of error for each identified so that they can be minimized.

LIMITING DILUTION

This is an alternative to the plate count, which may in theory be used for organisms which it may not be possible to grow as colonies (e.g. in the case of the highly motile spirochaete *Leptospira*) and where a viable count is essential (i.e. the alternative of, for example, a visual count is unacceptable). The principle is to perform serial dilutions, and from each to inoculate a series of replicates for liquid culture, perhaps in tubes. The replicates are incubated and observed for growth of the organism. At the point in the dilution series where there may, or may not, be an organism in the volume transferred to a replicate tube, not all the tubes will show growth. From the proportion of tubes showing growth at each dilution step, a value for the viable count can be estimated. Note however the word 'estimated': the method is actually extremely poor statistically compared with the plate count, and the count is sometimes referred to as the 'most probable number', reflecting its poor reliability. The best interpretation of the data obtained is achieved by the use of special tables, given for example by Meynell and Meynell (1970), along with an excellent commentary on the mathematics of such 'quantal' counting methods.

The poor performance of limiting dilution methods for viable counts as against plate methods is referred to above (Chapter 8). For practical reasons the number of replicate tubes used has to be limited, and cannot approach the number of 'slots' provided by an agar plate for occupation by a separate and countable colony – perhaps 1000 or more per plate (not all of which have to be 'occupied' in order to be useful – the empty spaces on an agar plate are in a sense conveying the same information as tubes without growth in a limiting dilution assay).

VISUAL COUNTS

When the absolute number of microbial particles, dead or alive, must be known the microscopic visual count is a useful method. It has the disadvantage that it is not very sensitive: minima of about $10^7\ ml^{-1}$ of bacteria, or $2 \times 10^6\ ml^{-1}$ of larger cells such as yeasts, are needed for statistically valid counting. The reason for the greater sensitivity with larger cells is that they can be counted in a haemocytometer, whereas smaller cells must be counted in a shallower bacterial counting chamber of the Thoma type (see Chapter 2). Both types of chamber have similar ruled patterns to delineate areas to be counted; the haemocytometer is sometimes half-silvered to provide enough contrast for visualization of cells without the use of phase-contrast optics (otherwise the substage condenser diaphragm must be closed down to induce contrast – the slide is too thick for use with phase contrast). The latter is essential for the counting of bacteria in a Thoma chamber.

USE OF COUNTING CHAMBERS

Correct use of counting chambers is essential if results are to be reasonably reliable. The two types are shown in Fig. 51. The first requirement with both types is absolute

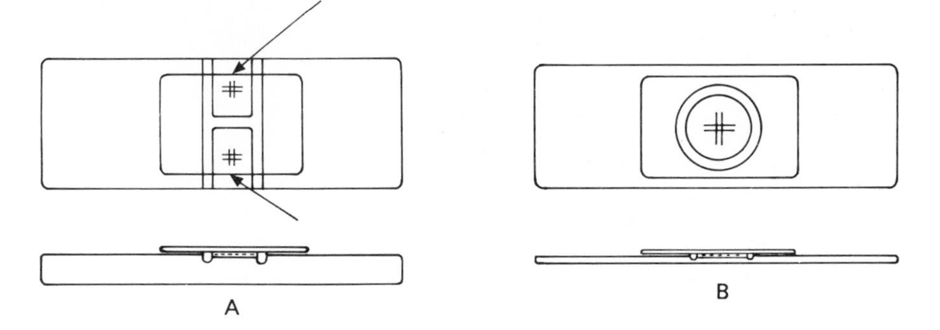

Fig. 51 Microscopic counting chambers. The haemocytometer (A) has a chamber depth of 0.1 mm and is filled by capillary movement of drops of liquid applied at the arrow points. The bacterial counting chamber or Thoma counter B has a chamber depth of 0.02 mm, and is filled by placing a few microlitres of liquid in the centre of the platform and lowering the coverslip on top.

cleanliness – they should be washed in detergent, but not soaked in it since the surface of the glass may be etched or half-silvering damaged or removed if present. After thorough rinsing and drying with a clean tissue, residual dust should be blown away. The haemocytometer is prepared by placing the coverslip in position, and the closeness of contact between the two glass surfaces can be checked by looking for 'Newton's rings', a variable curved pattern of parallel interference lines in spectral colours which will be visible if light is reflected from above off the surface between slide and coverslip. If surfaces are clean, these lines will form without the need to apply any pressure, and they are an indication that the surfaces are sufficiently close for accurate work. If dust is trapped between the surfaces preventing close contact, the rings will not be seen. It is sometimes said that firm pressure and slight movement from side to side should be applied to the coverslip so that close contact is ensured, but in my experience a class of students trying this can break up to half the special flat (and expensive!) coverslips. When proper contact between slide and coverslip is established, a small drop of the suspension to be examined is introduced beneath the coverslip by placing it with a pasteur or micropipette on the slide at the edge of the coverslip, so that capillarity will carry it across to fill the chamber. Just enough liquid to do so should be added.

The Thoma type of chamber is filled differently. Again glass surfaces must be clean, but prior contact before adding liquid is not possible. A small drop, about 2 mm across, of suspension is placed on the centre of the raised area of the slide, and the coverslip gently dropped onto it. Surface tension between the surfaces will pull the coverslip close to the slide, but it should be checked carefully to see that no liquid escapes from the gutter to run between the slide and coverslip outside the central raised area, and no dust is trapped in this outer area between the slide and coverslip.

When the slide has been prepared, cells should be allowed to settle for a minute or two before counting commences. Cells should not be so dense that they are difficult to distinguish individually, or so sparse that they are difficult to find. There is no fixed number of squares that should be counted, but ideally, if cell concentrations are appropriate, a distribution of squares from various parts of the slide (Fig. 52) should be counted. For statistical significance a minimum number of cells should be counted, in general at least 100 for approximate work and about 400 for the greatest accuracy. The

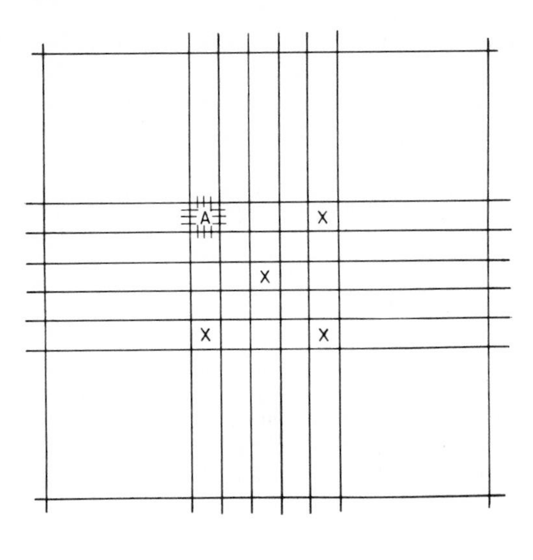

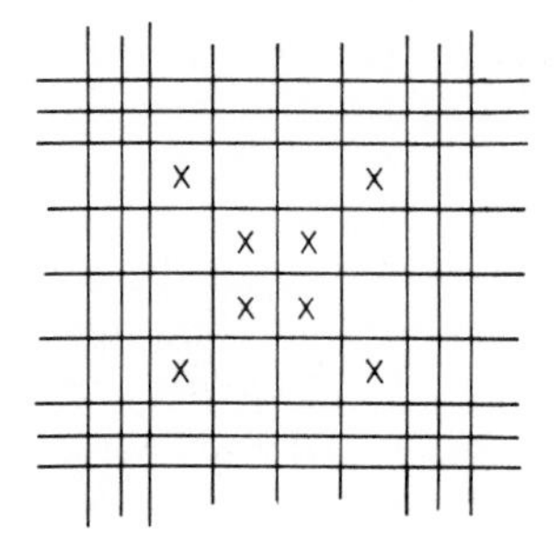

Fig. 52 The grid ruling on a counting chamber surface. The principal rulings are shown above, and the detail at A is enlarged below. The smallest squares shown are each $1/400\ mm^2$, and are arranged in 25 groups of 16, totalling $1\ mm^2$. To ensure unbiased sampling of squares for counting, patterns of squares to count may be identified as shown by the crosses if the numbers of organisms are too great to count the entire square millimetre.

grid pattern is shown in Fig. 52. For calculation of results, rather than trying to remember (possibly wrongly) a formula, it is easier to work out the factor to multiply the count by depending on the number of squares counted. What is needed first is the number of cells per mm^2; remember that each of the smallest squares is $1/400\ mm^2$, so the mean number per smallest square $\times 400$ is the number per mm^2. Then, multiply by 50 (or by 10 for the haemocytometer with its deeper chamber) to give the number per mm^3. This in fact is a measure commonly used by haematologists to report blood cell counts, but in microbiology numbers per ml (cm^3 or $10^3\ mm^3$) are more commonly used, so the count per mm^3 should be multiplied by 10^3. Again, any initial dilution factor should not be forgotten! Thus if, after an initial dilution of 1 in 5 a total of 50 smallest squares contained a total of 110 cells, we would have 880 or 8.8×10^2 in 400 smallest squares, i.e. $1\ mm^2$. This is $8.8 \times 10^2 \times 50 = 4.4 \times 10^4\ mm^{-3}$, or $4.4 \times 10^{-7}\ ml^{-1}$. Therefore before dilution, the total count would be 2.2×10^8 cells ml^{-1}.

Note that the minimum number of bacterial cells that is convenient to count accurately is about one per smallest square, or 400 mm^{-2}, which is 2×10^3 mm^{-3}, or 2×10^6 ml^{-1}. The accuracy of the method can be surprisingly poor. Errors due to variation in chamber depth, caused by dirt between glass surfaces or variation in the surface tension pulling the coverslip downwards, are difficult to eliminate. In addition, the statistical variability in the distribution of cells is considerable, and inconveniently large numbers of cells and squares may have to be counted for reliable results. Cells which clump in suspension are also very difficult to count by this method.

COULTER COUNTER

This is an alternative to the visual count, in that it counts all particles. The method is based on changes in electrical conductivity through, or across, a narrow orifice when cells in suspension are forced through it. Electronic circuitry enables the very rapid counting of electrical pulses generated when particles pass through the orifice, and because it is automatic the total numbers counted can be very large, improving statistical accuracy beyond that easily obtained by visual counting. A major drawback to the method is that the equipment is very expensive. It is mainly used in haematology.

OPTICAL METHODS: ABSORBANCE MEASUREMENT

Measurement of 'absorbance' of light by microbial suspensions is perhaps the most widely used and convenient method for quantitation. In fact, the physical basis for the method is scattering of light rather than absorbance, but the effect on light passing through a suspension is similar and it obeys the Beer–Lambert law of light absorbance over a limited range of cell densities. The law states that

$$\log_{10} \frac{I_0}{I} = Elc$$

where I_0 is the incident light intensity and I the intensity of transmitted light. Thus I_0/I is the ratio of incident light intensity to that transmitted by the specimen. $\log_{10} I_0/I$ is designated as absorbance; it is useful to remember that if 10 per cent of incident light passes through the suspension, absorbance will be $\log_{10} 10/1 = 1$; or if 1 per cent passes through, absorbance is $\log_{10} 100/1 = 2$ (in practice about the highest value of absorbance which can be measured meaningfully, and even this is likely to be in the non-linear part of the absorbance curve – see below). Light absorbance is conveniently measured with a spectrophotometer, commonly at a wavelength of 550 or 600 nm, or with a simple colorimeter with an orange or red filter. In the case of chemical substances which truly absorb light in spectroscopy, E is the molar extinction coefficient, a constant characteristic of a chemical substance absorbing light at a particular wavelength; therefore the amount of light absorbed is proportional to the path length l through the suspension, and the concentration c of the absorbing substance. In measuring cell densities, there will still be a constant equivalent to E which is characteristic of each cell type and size; in order to obtain absolute correlations of absorbance with cell number, the method would have to be calibrated for each cell type used against an absolute method such as the microscopic visual count. Thus absorbance will depend on the path length and cell concentration, and if measured for a fixed path length over a suitable range of cell concentrations is a direct

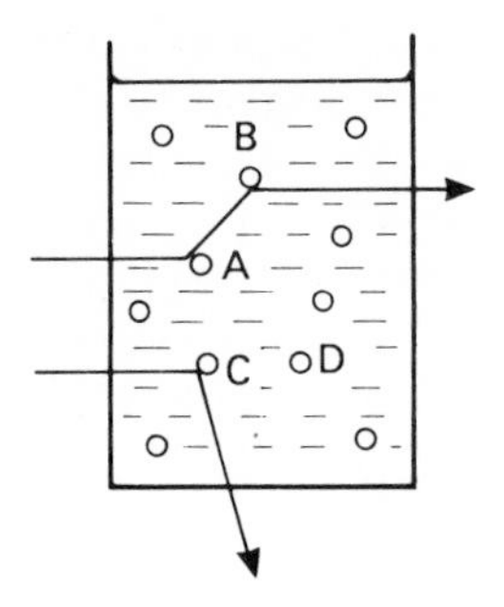

Fig. 53 Light scattering by particles in suspension. Effects which may distort the direct relationship between particle concentration and light scattering at high cell concentrations are shown. Secondary scattering may occur, as by cell B of light scattered by cell A so that it is not entirely lost from the incident beam. Alternatively, cell C may effectively shadow cell D, so that not all cells equally influence the degree of light scattering.

measure of cell concentration. The cell size factor means that the relationship between absorbance and cell number will not necessarily be constant at different phases of the growth curve, for example.

The limitations of absorbance measurement can be appreciated if we consider the consequences of reducing the cell concentration well below the number which will scatter, for example, 1 per cent of the incident light. This means we must detect differences of well below 1 per cent between the light intensities transmitted by the suspension and by the diluent alone; in practice colorimeters, or even sophisticated spectrophotometers, are unable to do this with accuracy. At the other extreme, if cell suspensions are at high enough concentrations to give absorbances higher than about 1, secondary light-scattering effects come into play and the absorbance curve loses linearity against concentration. These effects, of secondary scattering or of shadowing, are shown in Fig. 53.

For optimal performance of the method, absorbances should lie approximately in the range 0.1–1.0. This can be achieved by dilution, if necessary of concentrated suspensions, but if cells are few there is no easy way to measure absorbance. Thus the lower limit of accurate measurement is about 10^8 ml^{-1} for a bacterium such as *E. coli*, corresponding with an absorbance of about 0.1 at a pathlength of 1 cm. For yeast cells, being larger, the absorbance constant of the cells is considerably higher, and a minimum of about 2×10^6 ml^{-1} can be measured.

Errors in the measurement of absorbance can be minimal with careful dilution and attention to the factors described above: good spectrophotometers are precision instruments. The interpretation of results has to be careful however: dead cells, particulates other than cells, and solutes absorbing at the wavelength used will all erroneously increase the readings obtained. The first caveat may not be a problem though if we wish to measure the total number of cells, dead or alive! The other pitfalls can often be avoided by the use of appropriate blanks.

TURBIDITY MEASUREMENT

Turbidity measurement, known as nephelometry, is more sensitive than absorbance measurement and can therefore increase the useful range of accurate optical methods below that available from absorbance. The reason is clear: in the example above, the accurate measurement of 1 per cent decrease or less in the amount of light absorbed by a suspension was difficult; if however the light scattered is measured instead, it can be quantitated accurately and again is proportional to the cell concentration. Again however, the method becomes non-linear at high cell concentrations. In practice, since spectrophotometers are more widely available in laboratories and used for other purposes, transmitted rather than scattered light measurement is more commonly used.

DRY WEIGHT

Measurement of cellular dry weight is one form of absolute determination of biomass, which is independent of all cellular variables such as viability, cell size, etc. Hence it may be useful, for example in determination of biomass yields from different growth substrates. The method is inconvenient and slow however, since centrifuged or filtered and washed cells must be dried in an oven or desiccator to constant weight, and it is not very sensitive: to obtain the minimum of about 10 mg which can be weighed accurately on most laboratory fine balances, at least 10^{10} cells will be needed. In addition, there may be problems in removing all solutes from delicate cells without causing cell lysis and loss of cytoplasmic material, and particulates from the growth medium may be difficult to exclude.

CHEMICAL MEASUREMENT

A useful alternative to dry weight determination may be the chemical assay of a cellular constituent such as nitrogen, phosphate, protein, DNA or RNA, or a determination of biochemical activity. Examples of the last are luminescence measurement of adenosine triphosphate levels, or assay of an enzyme activity which may be proportional to cell number or mass. The experimental details of these types of assay are beyond the scope of this book, and the simple chemical determinations are widely described elsewhere. An advantage of chemical or enzymic assays is that they may be directly related to the parameter of interest in growing a microbial culture. If the optimum conditions for production of cellular mass on simple media were being investigated, determination of protein might yield directly the information required.

BIOLOGICAL ASSAY

In some circumstances, it may be necessary to assess growth in terms of the biological activity of organisms or their products; for example, the virulence or infectivity of organisms in a culture may vary independently of their viability, depending on growth conditions. For this and many other types of assay in biology, it is important to grasp the principles of bioassay. In many situations in biology, such tests have essential elements of dose and response: following the addition of a known quantity of an active agent into a system, a corresponding response is measured. Examples are the assay of an antibiotic or disinfectant, to assess its power to kill microorganisms, and the testing of a strain of bacteria or a microbial toxin for lethality towards animals or tissue culture cells. There are two variables in many such tests: the *dose* of an agent added to the system may be varied, and often the *response* of the system may be variable – there may be intermediate levels of response between complete killing of the target

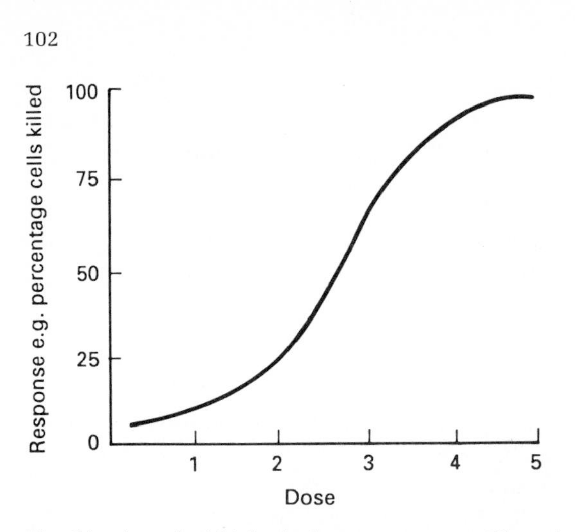

Fig. 54 A typical biological response curve. There is a good linear relationship between dose and response only over a rather narrow range of doses.

population, and complete survival. The design of such tests is often a crucial factor determining the value of the data obtained.

The most common error is to consider only the *response* in quantitative terms, and to use fixed dose levels of the agent being tested. For example, a bacterial toxin might be assayed in a cell culture system, and either cell death, shown by 'rounding up' and detachment of cells from culture vessel surfaces, or other 'cytopathic effects' such as changes in cell shape, might be assessed microscopically. Cells affected might be counted, and the percentage score compared between different toxin preparations: perhaps 50 per cent of cells will be affected by one preparation, and 75 per cent of cells by another. We may be tempted to assume that the second preparation is 50 per cent more potent than the first, since the effect is 50 per cent greater. Figure 54 shows however that this is a very dangerous assumption: if the dose used is varied and the response measured at a range of doses, the nature of the relationship between dose and response can be seen clearly, and there may not be a simple correlation of dose and response. In practice, the best way to handle such comparisons of potency of preparations in biological systems is often to identify a fixed level of response which is 50 per cent of the maximum that can be obtained, and to vary doses in order to find the dose capable of eliciting the 50 per cent end-point response. The advantage of this is evident from the graph: at this point, the dose–response curve is at its steepest – so the smallest variation in dose will elicit the maximum variation in response. Comparisons of doses which cause either minimal or maximal responses will not yield any information on their relative potencies.

CHAPTER 10

Harvesting and processing organisms

Although microorganisms may be grown for a wide variety of purposes, often the initial steps in harvesting and processing are similar, and they may usefully be summarized.

Harvesting

GROWTH FROM SOLID MEDIA

The simplest procedure for removal of growth from an agar plate is to use the loop as a miniature scraper. Many bacterial species yield material of quite a firm, cohesive texture on removal of colonies from agar, and a loop can be used to remove from agar a volume of packed cells equivalent to a small drop of liquid. Others may produce a viscous liquid from colonies, less easy to pick up, but with practice the loop is surprisingly effective. Material from plates may be suspended in liquid for further processing. If a relatively small amount of material is needed, it is usually best taken from areas of the plate which are not covered in confluent growth, for two reasons. First, if the plate is not too old, the cells in separate colonies should at least be mainly viable, even if many from the centres of colonies will be in the stationary phase. Second, it is often possible in areas of heavy confluent growth for small numbers of contaminating organisms from the inoculum to be present but unseen, whereas among separate colonies there would almost always be some difference in colony appearance of contaminants to alert the experimenter. Incidentally, growth from agar plate cultures can be sampled from plates several times, perhaps extending over several weeks after they have been incubated and then stored in the fridge. Often the properties of the cells will change rapidly on storage, and this practice should be avoided if reproducibility between experiments is necessary.

The scale of harvesting from agar plates is best increased with a scraper easily made from a pasteur pipette, by sealing the end and bending it at right angles in a bunsen flame (Fig. 55). The bent end can be made about 1 cm long for convenient small-scale suspension in a 'bijou' (5 ml) bottle, by vigorous agitation in diluent to suspend colony growth. Again, this is straightforward with firm, cohesive colonies but more awkward with semi-liquid growth, for which it is best to add a little liquid to the plate

Fig. 55 A simple scraper for harvesting agar-grown cells, made from a pasteur pipette with flame-sealed tip and end bent at right angles.

(uncontaminated plates only!) and suspend the growth with the help of the scraper, removing the liquid suspension with a pasteur pipette.

Sometimes it may be more convenient to use a cotton-wool swab to remove growth from plates, suspending it in diluent by rolling motions on the inside of the vessel, but this tends to be wasteful of material which will remain trapped in the cotton.

On a larger scale still, trays or flat 'roux' bottles of agar medium can be cultured, and larger scrapers used to remove growth. Often this would only be done with fastidious organisms difficult to culture in liquid. Large square pre-sterilized disposable plastic dishes about 25×25 cm are available.

To produce even suspensions of agar-grown organisms, necessary for quantitation whether optically or by particle count, simple agitation by vortex mixer may be sufficient. In some cases growth is too sticky to suspend easily, and addition of coarse glass beads of several millimetres in diameter before vortexing may help. It may be necessary to remove coarse particles, for example of agar medium, by filtration through a cotton-wool plug before further processing.

Care should be taken in selection of a diluent for suspending agar-grown cells. Very fragile organisms may need a medium with particular osmotic or ionic properties. For osmotic stability, saline or buffered saline should be the minimum. Often, fragile organisms, particularly Gram-negatives, may be stabilized by addition of divalent cations such as magnesium at a low concentration (e.g. 10 mM). Protein may also stabilize, and a low concentration of serum or a purified protein such as albumin may be used, or cells may be suspended in a simple broth medium.

Many bacterial species are straightforward to grow and harvest on solid media as described above. However, not all are so obliging and special procedures may have to be devised. Colonies of some species will not detach easily from agar surfaces; this applies especially to filamentous fungi and streptomycetes, for which special techniques have been developed. Some unicellular bacteria will not form 'smooth' suspensions of individual cells which can be diluted and counted or optically quantitated by the methods described above. With ingenuity alternative methods can be developed; cells can be pelleted by centrifugation and weighed wet, for example.

Technical inconvenience should not, but often does, deter scientists from working with difficult systems, and many of the most fascinating problems which need to be explored have technical barriers associated with them!

GROWTH FROM LIQUID MEDIA

The product of liquid cultures is already in a convenient form for processing, albeit with one major disadvantage in terms of large-scale production: large volumes of liquid are not easily handled, especially if cell densities are low. It is therefore advisable, where possible, to grow cultures to high densities if large amounts of cells are required. Also, if extracellular products secreted or lost from the surface of the organism into the growth medium are required, recovery may be even more problematical than for whole cells.

CENTRIFUGATION

Centrifugation is one of the most useful procedures available to the microbiologist for handling and processing microbial cells, both for harvesting them and for washing them free of unwanted solutes. Very high speeds or long runs are not usually required, and standard bench centrifuges generating accelerating forces of about 3000 ***g*** are adequate for many small-scale applications. Better still for small-scale work are microfuges, for reasons explained below. Centrifugation at its most sophisticated is a complex and expert science, and detailed advice can be found in specialist reference books (see Further Reading). For routine use to pellet cells, however, a few simple rules will help microbiologists to get the best from the centrifuge. Elementary physics tells us that the centrifugal acceleration on a particle spinning in a centrifuge is proportional to the radius of the circle in which it is spinning, and to the square of the angular velocity. The force on the particle is the product of this acceleration and its mass. Particles in a vacuum would immediately fly to the bottom of a centrifuge tube, but in practice two forces combine to restrict their movement – frictional force dependent on the viscosity of the suspending medium, and the buoyancy of the particles dependent on their density relative to that of the liquid surrounding them. In practice, for routine applications the last two are quite constant in aqueous solutions – most buffers and media of average physiological solute content have densities quite close to 1 and viscosities quite close to that of water. (One important exception is the high viscosity of high-molecular-weight DNA: if organisms lyse and release chromosomal DNA during processing, it may be very difficult to pellet them effectively.) Therefore the major variables normally are radius and velocity. One other indirect factor is important however: because microorganisms are so small and hence have a small mass, the centrifugal force acting on them is small and movement, even through low-viscosity, low-density media, is slow. Therefore, the shorter the distance they have to travel, the more rapid and efficient will be the centrifugation process. For another reason too, difficulties are caused if the distance to be travelled from top to bottom of the tube is large – there will be a significant difference in radius, and hence in centrifugal force applied, between the top and bottom of the tube. The whole centrifugation process may be significantly slower if the particles at the top of the tube have to migrate much more slowly at first, and then eventually accelerate to join those already pelleted at the bottom of the tube.

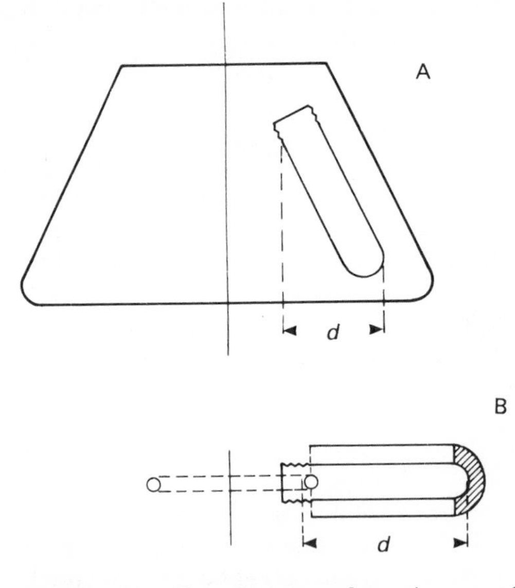

Fig. 56 The effect of rotor configuration on sedimentation efficiency. The horizontal distance *d* through which particles have to be moved from the top to the bottom of the tube for sedimentation is less for an angle rotor (A) than for the same sized tube in a swing-out rotor (**B**). Particle pelleting is therefore quicker, and the effect is more marked with angle rotors in which the tubes are closer to the vertical.

The density of bacterial cells is also crucial to the success of centrifugation as a means of harvesting them. Different classes of biological macromolecules, despite the fact that all are composed mainly of carbon, hydrogen, oxygen and nitrogen, tend to have different densities depending on their degree of hydration. Lipids tend to form relatively anhydrous hydrophobic domains, and contain high proportions of hydrogen and relatively little oxygen and hence are not very dense. Proteins tend to have high densities, often as much as 1.3 in aqueous solution. On the other hand, carbohydrate macromolecules are often highly dispersed in a gel-like form with considerable hydration and their density is closer to that of water. These differences in density are often exploited in centrifugation for separations based on differential buoyant densities (see below).

Choice of rotor Angle rotors have two major advantages. The distance travelled by particles is shorter for the same length of tube than in a swing-out rotor, as shown in Fig. 56, so the time needed for pelleting particles is reduced. The rotor will normally have smooth contours, so air resistance is reduced compared with a swing-out rotor and heat generated by friction against the air is less. The latter becomes a real problem with speeds above about 5000 r.p.m. in large centrifuges, which require refrigeration to counteract it. A disadvantage of angle rotors is that the pellet will be deposited on the side of the tube, and if it is not well compacted it may settle under gravity and be disturbed when the centrifuge comes to rest, whereas in a swing-out rotor there is no such disturbance. Large angle rotors for substantial volumes, i.e. for bottle sizes larger than about 300 ml, would be too massive to handle and to accelerate and decelerate in the centrifuge, so for this application swing-out rotors are normally used, in refrigerated centrifuges.

For routine benchtop centrifugation of 5–50 ml volumes of bacterial suspensions, either type of rotor is acceptable, although swing-out rotors are perhaps preferable for spinning culture bottles not designed as centrifuge tubes since the stresses on the bottle are aligned down its walls rather than across the bottle, and breakage is less likely. Universal 25-ml glass bottles can be spun safely at speeds up to about 2500 r.p.m., which is adequate for many bacterial cultures.

The advantages of the microfuge are mentioned in Chapter 2: small tubes require only short run times, so speeds may be high since the time required is not sufficient for significant heating to occur. The high speed more than compensates for the small radius (since acceleration is proportional to the square of velocity). In consequence these instruments are also very compact.

Safety and centrifugation The centrifuge is potentially one of the more hazardous instruments in routine use in many laboratories. Accelerations several thousand times that of gravity, attained even in the most basic benchtop machines, could produce lethal projectiles if rotors disintegrated. Great care and respect are therefore required in their use. Tubes must always be well balanced to avoid stresses on rotors and bearings. Tube caps must always be carefully checked, and in angle rotors tubes should not be filled above the point where liquid will be forced out around the seal by high ***g*** forces – see Fig. 57 which illustrates that during spinning in angle rotors with near-vertical pockets, the capacity of the tubes may be strictly limited if the risk of leakage is to be avoided. On high-performance centrifuges, however, tubes must be full to avoid their collapse and sophisticated sealing mechanisms may be used to prevent such leakage. Inevitably, leakage does often occur, both in angle rotors and into the buckets of swing-out rotors, which some workers deliberately use as reservoirs for water added to balance tubes before spinning. All such contact of rotors or buckets with liquid must be remedied by rinsing and drying them after use to prevent corrosion and the risk of consequent failure. Aluminium alloys, often used for angle rotor construction to avoid excessive weight, are especially liable to corrosion by inorganic salts present in all media and buffers, so the rinsing is particularly important.

Care must also be taken in opening centrifuges; modern instruments should have an interlock device to prevent opening of the lid before the rotor stops spinning, but on

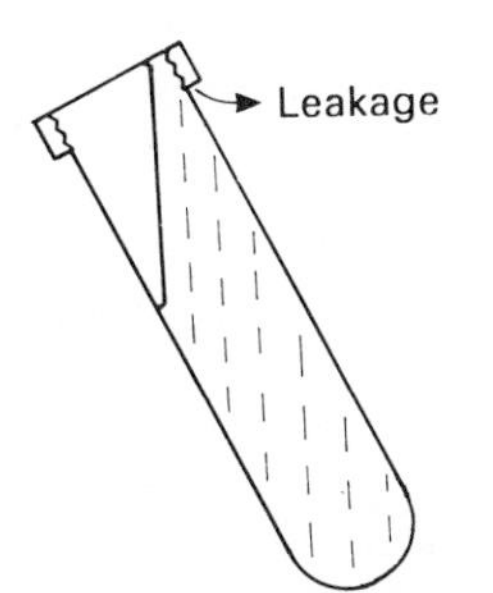

Fig. 57 A problem with angle rotors is the potential for leakage of contents between the tube and cap during the run, when the liquid surface is vertical and a high pressure is exerted by the centrifugal force on liquid 'above' the seal.

older machines this may not be present and the lid should not be opened prematurely – if clothing or hair became entangled in a spinning rotor, serious injury could result.

Most important also in the centrifugation of microbial cultures is the risk of generation of aerosols, a potential source of accidental infection. If drops of culture are carelessly allowed to remain on the outside of bottles or tubes, or if a container should fracture during spinning, the released suspension will be distributed very effectively as an aerosol in the highly disturbed air around the rotor. Therefore, in centrifuging any living organisms which carry a significant risk of infection by aerosol, sealed centrifuge buckets which can be obtained easily from manufacturers should be used.

Processing of cultured microorganisms

Microbial biomass is grown for a variety of purposes, many of which involve further processing to isolate fractions of interest, whether for their enzymic, toxic, antigenic or other biological activity. Often the initial steps in processing are common, and they depend on the cellular location of the fraction sought and the general nature of the organisms – particularly their Gram character. It may be helpful at this point to summarize the main structural features of Gram-positive and Gram-negative bacteria (Fig. 58). A major difference is in the thickness and complexity of the main structural layer, the peptidoglycan complex, which is thick and interwoven with other macromolecules, mainly carbohydrate, in the Gram-positives while it is essentially a thin molecular monolayer sacculus in Gram-negatives. The second essential difference is in the possession by Gram-negatives of an outer membrane, very different in composition and function from the cytoplasmic membranes of both major types.

CULTURE FILTRATE

Microbial products may accumulate during growth in the soluble fraction of a culture: examples are common among extracellular toxins and enzymes which are actively secreted. There are several simple ways to concentrate such products. The process will

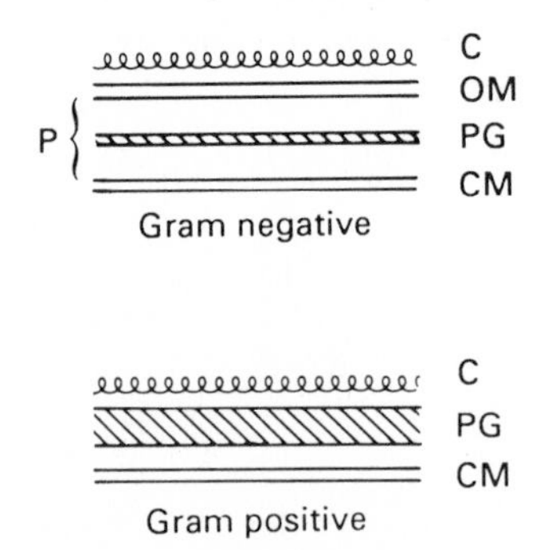

Fig. 58 The basic structural features of Gram-negative and Gram-positive cell walls. Note the additional (outer) membrane (OM) in the Gram-negative, enclosing the periplasm (P), and the thicker peptidoglycan (PG) layer in the Gram-positive cell. Both cell types, like all other living cells, have a cytoplasmic membrane (CM) and may also have capsules (C).

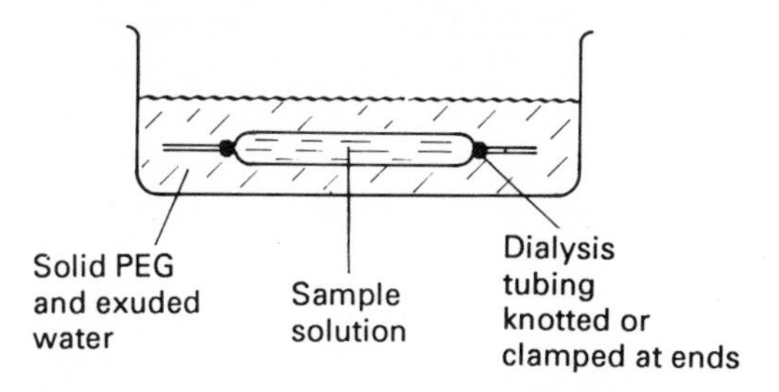

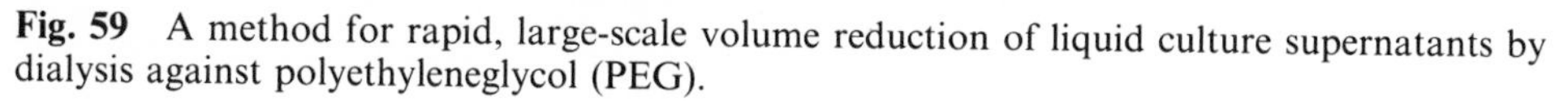

Fig. 59 A method for rapid, large-scale volume reduction of liquid culture supernatants by dialysis against polyethyleneglycol (PEG).

be most efficient if cultures are grown to the highest density compatible with prevention of degradation of the cells or their products due to autolytic processes or degradation by, for example, proteolysis if the organism produces potent proteases. In general the use of rich, well-buffered culture media and shaken liquid cultures taken to the end of the logarithmic growth phase will produce the highest yields in most concentrated form, but assays of activity must be done to ascertain the period of maximum production of the product of interest: secondary metabolites will usually be produced after logarithmic growth ceases.

Culture filtrates, as they are commonly known, or more usually supernatants from which cells have been removed by centrifugation, can be concentrated by several methods. Precipitation of macromolecular components by 'salting out' can be achieved by adding high concentrations of salts such as ammonium sulphate which interfere with the hydration of proteins, or methanol or acetone to achieve the same effect. Precipitation tends to be most effective at the isoelectric pH of the protein concerned, and in some cases isoelectric precipitation alone is effective. Another alternative is to reduce the water content of the culture supernatant by dialysis against high-molecular-weight polyethyleneglycol, which has a high affinity for water and will attract it through a semi-permeable cellophane dialysis membrane. The latter is available in tubular form (Visking tube) which can be tied off when wet to seal-in the contents. The polyethyleneglycol is unable, due to its large molecular size, to cross the membrane and enter the solution being concentrated (Fig. 59). Pressure ultrafiltration in a magnetically stirred pressure cell (Fig. 60), usually with pressure supplied by compressed air, is easily achieved for relatively small volumes, although large volumes may rapidly clog the membrane. These processes are quite slow and may allow time for significant degradation of unstable solutes to occur, in which case freeze-drying may be a useful alternative. It will have the disadvantage that all solutes, both low- and high-molecular weight, will be concentrated and upon re-dissolving freeze-dried material in a small volume the salt concentration may be extremely high. If the material sought is of high molecular weight, over about 10 000, prior dialysis against a large volume of distilled water before freeze-drying will alleviate the problem.

In some cases there may be significant levels of contaminating macromolecular solutes in the growth medium used, particularly in media for fastidious organisms. It may then be helpful to grow cultures in dialysis sacs within larger volumes of complex media, so that only the low-molecular-weight nutrients are able to penetrate and gain access to the microorganisms.

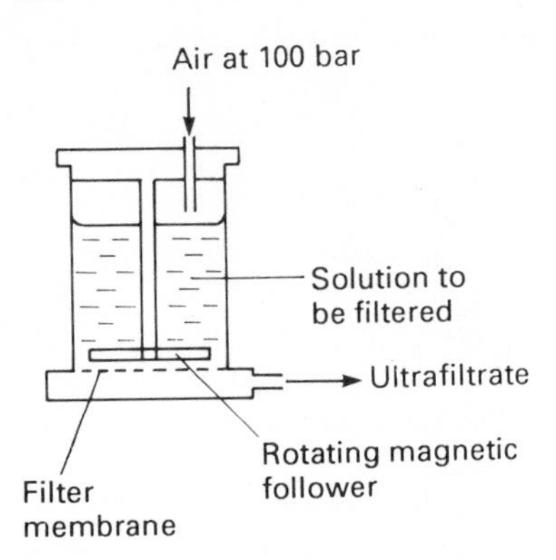

Fig. 60 An ultrafiltration pressure cell for concentration of liquids, including culture filtrates. The membrane is prevented from clogging by close proximity of a stirrer, which may be magnetically driven.

CELL-SURFACE COMPONENTS

Fractions enriched for cell-surface components can often be obtained by quite gentle solubilization procedures. Some organisms will release surface constitutents easily by gentle agitation of concentrated suspensions ($10^{10} - 10^{11}$ ml^{-1}) in physiological saline: Gram-negative *Neisseria*, for example, will yield substantial amounts of surface-derived vesicles of outer-membrane material this way, by vortexing a concentrated cell suspension and removing cells by centrifugation. More vigorous agitation in a blender, or mild ultrasonication, such as can be achieved in a bath rather than with a probe, may be more effective, and surface-located filamentous structures like flagella or fimbriae may be sheared off the cell effectively. Treatment of intact cells with high concentrations of salts such as potassium thiocyanate, or even sodium chloride at about 1 M, which weaken non-covalent bonding between surface macromolecules, may also enhance their solubilization. Chelating agents such as EDTA, used at about 10 mM, will remove divalent cations and reduce ionic stabilization, especially of the outer membranes of Gram-negative bacteria. Reducing agents such as 2-mercaptoethanol will break disulphide bonds which may stabilize high-molecular-weight aggregates of cysteine-rich surface proteins – a useful step in the preparation of yeast protoplasts and potentially valuable for enrichment of cysteine-rich outer-membrane proteins from Gram-negative bacteria. Detergents, often non-ionic types such as Triton, if necessary in combination with EDTA, may be used to release integral membrane proteins from the outer membranes of many Gram-negative bacteria, although such drastic treatment may lead to severe disruption of cellular integrity and release unwanted cytoplasmic components as well.

CELL BREAKAGE

Often it is necessary to break cells open physically in order to obtain surface structures or intracellular constitutents in quantity. There are several methods, both physical and chemical or enzymic, for achieving this. Classical methods include the use of presses such as the French press (see Chapter 2), or high-speed shaking or homogenization with very small glass beads of < 1 mm diameter (ballotini). An effective alternative is ultrasonication, which if administered by probe can be a very efficient method for breaking cells physically, especially for Gram-negative cells which are generally less robust than Gram-positives. The latter especially may require enzymic attack, for example by lysozyme, to weaken the cell walls before breakage.

FRACTIONATION OF CELL COMPONENTS

The details of subcellular fractionation are complex and highly specialized. The following notes give only an indication of the main possibilities.

Cell envelopes, i.e. walls complete with cytoplasmic membrane and, in the case of Gram-negative bacteria the outer membrane, are relatively easy to obtain from physically broken preparations of whole cells by centrifugation. Because the mass of the isolated cell envelope is substantially less than that of the intact cell, greater ***g*** forces are necessary to pellet it. Since we can rarely expect complete cell breakage, it may be advantageous to spin first at the minimum speed and duration necessary to remove whole cells, although at some risk of losing a proportion of the cell walls present. To pellet isolated envelopes, conventionally a refrigerated centrifuge able to generate speeds of about 20 000 r.p.m. (up to 45 000 ***g***) is used with run times of an hour or so; the microfuge provides a useful alternative, however, and 10 min at *ca* 12000 r.p.m. may be adequate. Washing pelleted walls several times by centrifugation may be necessary.

Cell-wall peptidoglycan may be obtained in quite pure form from crude envelopes by harsh procedures which remove all protein and lipid, but to which the backbone structure is resistant. Exposure to sodium dodecyl sulphate or aqueous phenol at high concentrations, or alternatively enzymatic digestion with proteases may be effective. Similar harsh degradative procedures may also isolate other fractions which are essentially carbohydrate. Capsular polysaccharides may be de-proteinized, and lipopolysaccharide obtained from whole cells of Gram-negatives by hot phenol–water extraction.

Cell envelope fractions obtained by physical breakage should also contain both cytoplasmic membranes and, in the case of Gram-negative bacteria, outer membranes. While these membranes are normally strongly associated with the peptidoglycan backbone layer of the cell wall, in the case of Gram-negatives by covalent bonding to certain outer-membrane proteins, the peptidoglycan can be degraded enzymically by digestion with lysozyme for example. This reduces the cell-wall polymers to relatively low-molecular-weight components, and the membranes remain as the main particulate fraction, again being obtainable by centrifugation which may need to be at a higher ***g*** force than above although again the microfuge can be surprisingly effective.

Because of their very different structural roles, locations in the microorganism and composition, it is not only of interest but also quite straightforward to separate cytoplasmic and outer membranes of Gram-negative bacteria. The classical method described earliest is based on the differential buoyant densities of the two membranes, exploited in ultracentrifugation. The cytoplasmic membranes of bacteria are essentially phospholipid lipid bilayers with integral membrane proteins. In the outer membrane, however, lipopolysaccharide, substantially denser than phospholipid, usually replaces much of the phospholipid in the outer leaflet of the membrane and as a result the outer membrane tends to be denser than the cytoplasmic membrane. This property makes possible the separation of the two membrane fractions by isopycnic density-gradient ultracentrifugation, often performed on sucrose gradients. An alternative method which is widely used, because it is much simpler, is to use detergents for differential solubilization of the cytoplasmic membrane in cell envelope preparations: outer membranes tend to remain intact as shown by their protein content, but some removal of phospholipid or lipopolysaccharide may occur. Such treatment of crude envelope fractions of Gram-negative bacteria will yield essentially peptidoglycan sacculi with attached outer membrane. The commonly used technique of sodium dodecyl sulphate–polyacrylamide gel electrophoresis (SDS–PAGE) analysis of such fractions will solubilize most of the protein of the outer membrane, whereas the peptidoglycan sacculus remains intact and does not enter the gel – it is easy to forget that it is still present in such preparations!

Many of the above methodologies are described in detail in 'Bacterial Cell Surface Techniques' – see Further Reading.

CHAPTER 11

Tissue culture

Tissue culture, while not a microbiological technique in that it is not concerned directly with microorganisms, has a great deal of overlap with microbiology in the practical approaches used, and indeed is frequently used in microbiology as an essential element in experimental work. It therefore seems appropriate to cover the general principles here.

Like microbiology itself, tissue culture can seem highly specialized and the exclusive province of trained experts, which is off-putting even to experienced microbiologists if they have never used it. This is quite unnecessary – with good aseptic technique and an understanding of the basic biology of cultured cells they are as easy to grow as most bacteria and easier than the most fastidious species!

CELLS AND CELL LINES

Most mature cells from organized tissues are terminally differentiated and are therefore incapable of division, although many can be maintained in a viable state for days or weeks. Those cells which will multiply in culture comprise two basic categories, *finite lines* and *established* or *transformed cell lines*. In addition *primary cells* are used, and are essentially those cells present in tissue sources which are able to survive in a viable and metabolically active state after disruption of the parent tissue.

Tissues which are rich in rapidly dividing cells are particularly good sources of primary cells, and include for example embryonic or fetal tissues; when cells from such sources which are capable of division are subcultured, they are regarded as cell lines but will not necessarily become continuous lines. Unfortunately, whatever the tissue of origin, such cells do not always have differentiated characteristics of that tissue, and many will be cells of fibroblast type which are preferentially able to divide. Frustratingly for biologists who may have taken a lot of trouble to establish continuous cell cultures from tissues, such cells do not continue to multiply in culture indefinitely, and after typically about 40 serial passages *in vitro* they may die out.

Transformed cell lines, however, have undergone a change in their basic character, for example in chromosome number, and in their regulatory mechanisms which allow them to multiply indefinitely. The change is known as transformation, and it seems to be analogous (although probably not identical in all ways) to the process which leads to formation of a cancer cell line in the body: many cancer cells in fact behave essentially as transformed cell lines in culture *in vitro*. Certain agents including some viruses and mutagens associated with causation of cancer can induce transformation among primary cells *in vitro*, again suggesting a link between carcinogenesis and cell transformation. However they arise and by whatever mechanism, which is only now beginning to be understood, transformed cell lines are extremely useful to the biologist and many have been used as standard tools over many years in many types of experimental system.

One problem which remains severe in many fields of experimental work with cultured cells is that in most cases they bear little real resemblance to differentiated cells which are the real foci of interest in considering, for example, interactions with infectious microorganisms. However, progress is being made, and methods for handling, for example, cells of the immune system, hepatocytes and mucosal epithelial cells are all becoming easier. Pre-eminent among these developments in the last decade has been the development of technology for 'immortalization' of specific antibody-producing cell lines of the immune system by their hybridization with related transformed cell lines derived from cancers of the lymphoid tissues: monoclonal antibody-producing lines, or 'hybridomas'.

A wide variety of well-known cell lines can be bought from major manufacturers of media and culture vessels and equipment. Many are also available from reference collections, which may be associated with microbial culture collections.

CULTURE VESSELS

Although cells can be cultured adequately in glass bottles, most work is now done in pre-sterilized disposable plastic containers. These range from microtitre plates with wells holding just 0.25 ml of culture fluid, through larger multiwell plates with 2- or 5-cm diameter wells, to flat flasks of surface area from 25 cm^2 to 175 cm^2. The plastics are specially treated and supplied for tissue-culture work, to ensure that surfaces are compatible with colonization by sensitive cell lines. Their use, while apparently expensive in terms of consumables costs, is justified by their greater uniformity and reliability than that of recyled glass; loss of perhaps several months' worth of work through unreliability of glassware is ultimately a lot more expensive! Microtitre trays for tissue-culture work often have deep divisions between the wells to help prevent the spread of microbial contamination from well to well.

LABORATORY FACILITIES

Most of the items required for tissue-culture work are similar to the basic requirements for microbiology, in terms of laboratory space and general equipment. Specialized items include a clean or sterile air environment (cultures are slow growing and easily outgrown by microorganisms, so exceptional sterile technique is paramount), an inverted microscope for inspection of growing cultures, and ideally an incubator providing a carbon dioxide-enriched atmosphere for buffering of bicarbonate-containing media (see below).

MEDIA

Tissue-culture media for mammalian cells are mainly based on rather empirically derived mixtures of amino acids, salts, an energy source such as glucose, and growth

Table 10 Several media for use in cell culture, and typical uses.

Medium	*Main components*	*Uses*
Minimum essential medium (Eagle) with Earle's salts	13 amino acids, 8 vitamins and growth factors, glucose, salts including $NaHCO_3$	Basic medium for a variety of primary and established cell lines
Dulbecco's modification of Eagle's medium	16 amino acids, many at double minimal strength, 8 vitamins and growth factors, glucose at 3 × minimal strength, $NaHCO_3$ at 2 × minimal	Higher yield, richer medium suitable for many common cell lines
Medium 199	21 amino acids, similar strength to MEM; 18 vitamins and growth factors, salts, 9 nucleic acid precursors, lipids etc., glucose and $NaHCO_3$ at MEM strength	A high-quality general-purpose medium
RPMI 1640	20 amino acids, 11 vitamins and growth factors, salts, glucose at double minimal level, $NaHCO_3$ at minimal level	Originally developed for leukaemia cells; often used for monoclonal antibody work

factors, to resemble in many parameters the composition of body fluids. The exact details of composition are not usually important to the user, since nearly all media are bought ready prepared from one of several major manufacturers. They can be obtained either as sterile, single-strength medium ready to use apart from addition of supplements, or as a tenfold concentrate, or as dried powder. If concentrates or powder are used, the water for dilution or dissolving the powder must be of the highest quality, for example distilled, deionized and filtered, and tested for low conductivity; if there is any uncertainty about it, suitable water can be bought or the single-strength medium should be used. The liquid forms can be stored for some months at ordinary refrigerator temperatures. Some well-known types are listed in Table 10, with some indications of their uses. The basal media are supplemented before use with several additional solutions. If buffered with sodium bicarbonate (see below), a concentrated solution of it is added before inoculation to give the correct initial pH of 7.4. Liquid media may need to be supplemented with L-glutamine, which is less stable on storage in solution than other amino acids. Serum is commonly added to media, most commonly to 10 per cent by volume; it supplies a variety of growth factors, hormones, proteins which mediate attachment to surfaces, lipids and other metabolites such as pyruvate which may be essential for some cells. Often fetal calf serum is used, because it is free from antibody molecules which may interfere with cell lines or viruses grown in them: the bovine placenta is impermeable to antibodies from the blood of the mother. Fetal calf serum is very expensive (the price fluctuates but is usually of the order of £100 per half litre), and other sera may be equally effective as growth supplements but may not be acceptable for other reasons, depending on the use to which the cells will be put. Serum proteins may in any case be a nuisance to the experimenter if, for example, soluble cell products are to be collected, and serum-free media have been developed which will support some, but not all, cell lines or types.

There is increasing recognition of a variety of growth factors which may encourage the growth of particular types of cell, especially primary cells, and the culture conditions which may influence differentiation, or survival of already differentiated cells, are becoming better understood. Many factors, some probably not yet recognized, are probably present in serum, hence the need to include it so often in media.

Finally, antibiotics are often added to tissue-culture media, to help maintain sterility by killing accidental contaminants. The need for them illustrates the seriousness of the sterility problem mentioned above: because cultured cells grow so slowly, and are kept often for weeks or months, they are outgrown by contaminants (in an ideal growth environment) far more rapidly than would be most microbial cultures. Commonly, benzyl penicillin (100 units ml^{-1}) and streptomycin (100 $\mu g\, ml^{-1}$) are used to provide a reasonable coverage against both Gram-positive and Gram-negative bacteria. Although there are potentially problems of resistance of environmental organisms to penicillin, more sophisticated alternative antibiotics do not seem to be widely used, perhaps because the toxicity of penicillin and streptomycin are exceptionally low. Sometimes however antifungal antibiotics, particularly amphotericin B (fungizone) are useful in combatting persistent fungal contamination, but with attendant problems of toxicity for cell cultures: the cytotoxic level of about 30 $\mu g\, ml^{-1}$ is not very far above the recommended antifungal level of 2.5 $\mu g\, ml^{-1}$.

BUFFERING OF TISSUE CULTURE MEDIA

The classical buffer used in tissue culture is the sodium carbonate–bicarbonate system, originally devised to mimic carbon dioxide–bicarbonate-mediated stabilization of pH in the blood stream. In tissue culture as in the tissues themselves, the main need for a high buffering capacity is to counteract the tendency for pH to fall as organic acids are produced by catabolism of glucose during cell growth. Some media, such as Eagle's medium, have a higher buffering capacity than others, and are hence able to support a higher concentration of cells.

The basic mechanism of action of bicarbonate is that in a regulated gaseous environment at a raised CO_2 level, normally about 5 per cent, any overproduction of hydrogen ions in the medium will push the equilibrium in equation (1) to the right, towards carbon dioxide, which can be lost to the system by diffusion out of solution into the gas phase.

$$H^+ + HCO_3^- \rightleftharpoons CO_2 + H_2O \quad (1)$$

There is however a major disadvantage with this system; if media are exposed to the normal atmosphere, in which CO_2 levels are very much lower, gaseous CO_2 is lost by diffusion in any case from a sodium bicarbonate solution and the pH rises rapidly due essentially to formation of sodium hydroxide as in equation (2).

$$Na^+ + HCO_3^- \rightleftharpoons Na^+ + OH^- + CO_2$$

Thus exposure to air of media buffered by reasonably effective concentrations of sodium bicarbonate leads to a rapid rise of pH, and this may be quite toxic to certain types of cell. Nevertheless, bicarbonate is widely used in cell culture, not least because it seems to enhance growth of many types of cell, for reasons which are not always exactly clear. There are however alternatives; commonly used laboratory buffers such as phosphate are not useful since the phosphate will interfere with metabolic processes, but a number of complex organic buffers have been devised which are metabolically quite neutral. The best known and most used is HEPES, a zwitterionic piperazine derivative which is used at concentrations of 10–20 mM, brought to the desired starting pH with NaOH. It does have some toxic effects for some cells above about 40 mM however. A low level of bicarbonate (e.g. 5–10 mM) should also be added to HEPES-buffered media as a nutrient factor. In the context of buffering, it should also be remembered that the complex mixture of amino acids and proteins in tissue-culture media also has a substantial buffering effect.

A vital safeguard in the use of cell-culture media, and indeed a very useful tool for monitoring the state of media during culture incubations, is the use of the pH indicator phenol red, which changes colour over the range from about 5.5 (bright yellow) to 8.0 (vivid magenta). At the desirable pH of 7.4, phenol red is a brick red or vermilion colour, and bicarbonate buffered media rapidly develop the tell-tale magenta colour if exposed too long to air. When cells are growing, acid production due to fermentation of the medium components is equally indicated by a colour change to orange and eventually to yellow as the buffer becomes exhausted. This is almost as good a guide to growth as microscopic examination.

PROCEDURES FOR CELL CULTURE

Media should be prepared complete and ideally the whole or a sample pre-incubated for several hours or overnight to check sterility (but bear in mind that some consituents, such as penicillin G, are quite unstable at neutral pH and medium should not be held any longer than this before use). Cells are initially seeded into fresh medium, normally at concentrations of 5–20 per cent of the maximum that can be supported before further subculture or 'splitting', and with several millimetres depth of medium above them. Incubation is at 37°C for 2–3 days with daily inspection to monitor for contamination, and to assess growth of cells. Most primary and many transformed cell lines grow attached to the surface of the glass or plastic vessel in which they are growing, and when attached and spread, dividing cells briefly exhibit mitotic figures which should not be difficult to find (although they will only be present in a minority of cells at any one time – see below) by phase contrast if an inverted microscope with this facility is available. If cells remain rounded up when expected to attach – which normally occurs within an hour or two of seeding – this may be an indication of poor viability. Non-attaching cells such as myeloma cells and hybridomas are different – see below.

Over a period of several days, depending on the doubling time of the cells and their initial density, the culture will approach confluence, i.e. cell progeny migrate to fill vacant spaces on the surface and adjacent cells come into contact so that a monolayer is formed. For reasons which are not fully understood, this cell-to-cell contact usually inhibits further growth even if the medium is not yet exhausted. Doubling times are commonly of the order of 12–48 h, and seldom < 12 h, hence the danger of overgrowth by contaminating microorganisms. The growth curve of cultured cells is in other respects however much like that of microorganisms, with lag, logarithmic and stationary phases. During logarithmic growth, increases of five- to twentyfold fold in cell number will take at least 24–48 h, even for the most rapidly growing cells.

When growth is complete, cells must be removed from the vessel surface for further culture or use. This is normally done by enzymic digestion of the adhesion proteins which mediate interaction with the surface of the vessel. Sterile trypsin solution with EDTA is supplied for the purpose by the main manufacturers of tissue-culture supplies, and used at 0.05 per cent (w/v) trypsin, and 0.02 per cent (w/v) EDTA. Spent medium must be removed first, and any remaining rinsed away with a serum-free balanced salts solution to remove any trypsin inhibitor (antitrypsin) which is present in serum: if cells do not detach within about 10 min, presence of inhibitor should be suspected. The trypsin solution is added to culture vessels in a volume just sufficient to cover the monolayer. After cells have detached they are often clumped together, and may be evenly suspended in fresh medium by gently sucking into, and expelling from, a pasteur pipette several times, before counting viable cells in a haemocytometer (see Chapter 9) and re-seeding fresh medium. Cell viability is assessed by the inclusion in the diluent of a dye which is taken up by dead cells but not by living cells. Traditionally trypan blue is used, but there is suspicion that it may be carcinogenic, and eosin is an equally good alternative, used at a final concentration of $0.4\,\mathrm{mg\,ml^{-1}}$. If serum-containing medium is to be used for subculture, there is no need to remove the trypsin by centrifugation and washing because the serum antitrypsin will inactivate it.

SUSPENSION CULTURES

Cell lines which grow in suspension are in some ways easier to handle, but it may be more difficult to judge when they should be split. Decreasing pH is an indicator of the need to change the medium or set up new cultures, but ideally a growth curve should be determined by counting cells at intervals, so that the period of logarithmic growth can be identified. Again, a haemocytometer is used and viability can be assessed by dye exclusion. Cells in suspension are less easy to check in the culture vessel by their microscopic appearance than those in monolayer, because the optical properties of the suspended cell do not easily allow visualization of internal detail – it acts like a lens and refracts light, rather than being spread on the surface in a thin layer through which light can pass. Nevertheless, the inverted microscope remains useful, and it is possible to distinguish viable and dividing cells by their bright refractility, large size, and lack of granularity. Dead cells in suspension tend to become less refractile, greyer and granular.

Many hybridomas, unlike some robust cell lines, seem to have a requirement for bicarbonate-buffered media, during the early stages after cell fusion. The recognition of initial colonies of hybridoma cells after fusion is a critical step for the novice, and is not always well described in textbooks. The cells usually have the characteristic bright refractility and roundness of suspension-growing cells, but characteristically cluster together, often at the edge of a culture well. They can be maintained for some days as colonies, but the most critical step in their development is subculture into larger volumes. The above comments about monitoring pH by judgement of colour, and viability by microscopic appearance, are particularly important at this stage, and all medium components, especially serum, should be the best available. Serum selected for efficacy in hybridoma work is commercially available.

MICROBIAL AND OTHER CONTAMINATION

The use of antibiotics can actually disguise sub-apparent microbial contamination, which must be checked for from time to time by plating samples on bacteriological media. A major hazard however arises from accidental contamination with mycoplasmas, which can coexist surprisingly well with tissue-culture cells, apparently as intracellular parasites. Unfortunately, none of the tests available for this contaminant is simple; the organisms are difficult to grow reliably, and the results of staining, for example with the Hoechst fluorescent dye 33258 followed by fluorescence microscopy, are not easy to interpret. Probably the least equivocal method is electron microscopy of thin sections of cells. The risk does, however, underline the need for extreme care to avoid contamination in the handling of cultured cells. Another interesting aspect of this need for care, with cell lines which may be subcultured subsequently for years or decades, is the reported presence in stocks of a variety of human cell lines of chromosomal markers characteristic of HeLa cells, thus suggesting that cross-contamination of stocks has taken place at some time in the past.

PRESERVATION AND STORAGE

Cultured cells can be preserved by freezing, provided precautions are taken to prevent the formation of highly structured ice crystals which will damage the cells. This is achieved by the use of cryoprotectants, solutes which can be present in quite high concentrations without causing great ionic or osmotic stresses for the cells. High protein concentrations are quite effective, and freezing in serum can be successful, but more reliable results are obtained with glycerol or dimethyl sulphoxide at 10 per cent by volume. In addition, cells should be frozen slowly, at a rate of about 1°C min^{-1}. Devices which achieve this, using liquid nitrogen or solid carbon dioxide as a refrigerant, are available (Fig. 61). Alternatively, use of an insulated container such as an expanded polystyrene foam box in a freezer at −70°C can be effective.

The procedure is to grow cells to mid- or late-log phase, when they will be in prime condition (do not use old cultures), remove if necessary from the surface by trypsinization, spin (low speed – a few hundred r.p.m. on a benchtop centrifuge) and resuspend in fresh medium with added cryoprotectant to 10^6–10^7 ml^{-1}, depending on

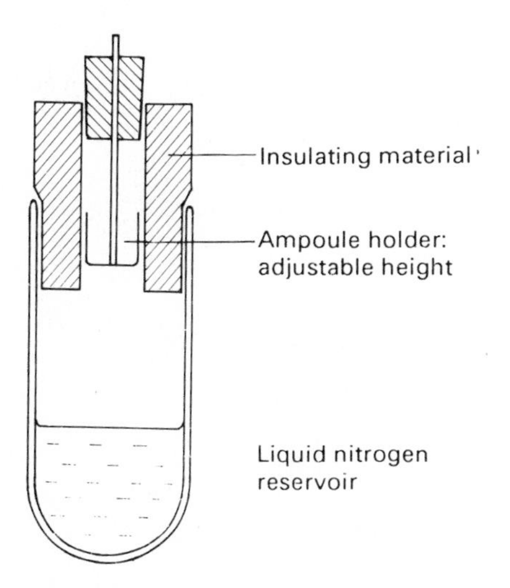

Fig. 61 A device for controlling the speed of freezing cells, by modulating the extent of thermal contact between the cryogen and cell suspension.

the reliability of recovery – i.e. robustness of the cells. Dispense in 1-ml volumes into ready labelled ampoules – pre-sterilized polypropylene screw-top 2-ml ampoules with silicone rubber seals are available – and commence freezing immediately. When cells are at −50°C or lower, transfer to permanent storage. Ideally, and essentially for long-term storage of more than a few weeks, this should be in liquid-nitrogen refrigerators, either in the liquid or, more safely, in the gas phase. The safety factor of the latter is that leakage of liquid nitrogen into ampoules is prevented. If it does occur, rapid boiling of liquid nitrogen on being immediately exposed to ambient temperatures after removal from the freezer can cause increases of pressure in the ampoule great enough to explode it violently. As a precaution a face visor should always be worn when removing ampoules stored in the liquid phase; the plastic ampoules can be inverted on removal so that any liquid nitrogen inside them is forced out past the seal by the pressure, thus rapidly removing it as liquid rather than allowing all of it to boil away to vapour. For short-term storage, −70°C freezers are adequate; freezing at −20°C is not effective.

RECOVERY FROM FREEZING

Ampoules removed from storage should be thawed rapidly at 37°C in a water bath (do not immerse the top of the tube – capillary action may carry the water-bath water, a notorious source of contamination, into the cap thread), with agitation to mix the contents as they thaw. The outside of the ampoule should be wiped thoroughly with 70 per cent ethanol to sterilize it before unscrewing the cap. The cell suspension should then be diluted about tenfold with complete growth medium; it may be advantageous with delicate cells to add medium gradually so that there is not a sudden change of solute environment. Ideally, cells should be centrifuged at about 200 ***g*** for 10 min on a bench centrifuge, and resuspended in fresh medium to eliminate the cryoprotectant, but again delicate cells may be harmed and it may be preferable to incubate immediately.

CHAPTER 12

Molecular microbiology – immunological and molecular biological aspects

One of the earliest applications of molecular analysis and manipulation to be developed in microbiology was in the study of immunological reactions of microorganisms with the infected or immune host: serology, the study of antibody-mediated phenomena, developed alongside microbiology at the time when a major spur to progress in both sciences was the successful treatment, therapy and prophylaxis of serious bacterial infections such as diphtheria and tetanus. Some knowledge of the basic procedures of serology as applied in microbiology is still a central part of a broad introduction to the subject. Genetic manipulation techniques, discovered through experimentation with microbiological systems, are being used increasingly by biologists in other disciplines, as their power for isolating, identifying and detecting genes and producing their products in microbial systems becomes known. Most of the microbiological manipulations of molecular biology are no different from those already described in this book, or are common-sense adaptations of them. There is, however, a set of more or less standard procedures which are commonly used, and an outline description of these will be helpful for the newcomer to microbiology, if only to make it clear how much can be attempted with quite ordinary facilities. Certainly there is no question of reproducing or substituting for the many excellent and comprehensive manuals which cover work of this type, some of which are listed in the Further Reading section.

Serology

ANTISERA

A wide variety of antisera for microbiological applications can be bought from commercial sources, for example for detection of organisms in clinical specimens or for typing isolates of clinically important bacteria, or as standard reagents for indirect or sandwich antibody-detection methods.Nevertheless, it may be necessary in a research laboratory to raise specific antisera from time to time. Facilities for this, the methods used, and the personnel approved to do the work, are all strictly regulated by the Home

Office under the *Animals (Scientific Procedures) Act (1986)*. Applications for both personal licences, for every operator who is to do the animal experimentation and who has to show evidence of training under experienced supervision, and project licences to show that the work proposed is valuable and justifiable and the methods appropriate, must be made in advance. The premises used must also be approved.

Procedures for production of antisera are quite standard, although a wide variety of optional techniques have been described for particular applications. For surface antigens of many bacteria, provided they are not toxin-producing strains which may be acutely harmful for laboratory animals, good antisera can be raised in rabbits by repeated intravenous injections twice weekly, of increasing doses of killed organisms (e.g. by exposure to 0.25 per cent formalin overnight) rising from about 10^7 to 5×10^8 organisms over about three weeks. A week later, the rabbit should be test-bled and should show a good titre of agglutinating antibody against the organisms. For protein antigens, or less antigenically potent organisms, the vaccine can be prepared by emulsification of the antigen (e.g. about $10^8\,ml^{-1}$) in mineral oil with a surfactant, either with (complete) or without (incomplete) additional killed organisms of *Mycobacterium tuberculosis*, i.e. the two versions of Freund's adjuvant. Emulsion vaccines should be administered intramuscularly or subcutaneously, in doses of perhaps 0.5 ml spread between several sites, twice or several times several weeks apart (a second or subsequent injection should always be administered in incomplete adjuvant, since harmful hypersensitivity reactions may develop to the mycobacteria), and the animals monitored by test-bleeds for development of the antibody response sought.

FLOCCULATION AND PRECIPITATION

Tests for immunological detection and quantitation of soluble antigen of microbial origin include the classical technique of the flocculation or precipitin test. This relies on the fact that at optimum proportions of antigen to antibody concentration in solution (provided both are high enough), an insoluble complex or precipitate, often with the appearance of floccules, will form rapidly (e.g. within minutes or an hour or two), upon incubation. When the proportions of the reagents are not optimal, the time taken is longer and the amount of precipitate formed is less. Although formerly used extensively, particularly for soluble antigens such as the toxins of diphtheria or tetanus organisms which are excreted into the culture supernatant, the method has now been largely superseded.

AGGLUTINATION TESTS

Although used less and less as new methods become more important for analysing the finer detail of antibody reactions with bacteria, agglutination tests remain the standard method for identifying bacteria at serotype level, particularly among the enteric group (see Chapter 13). The phenomenon occurs because most antibody molecules are divalent or polyvalent: they have two or more identical antigen-binding domains, and are therefore able to form a physical link or bridge between two antigenic particles, such as microbial cells (Fig. 62).

The procedure is simple for a *slide agglutination test:* bacteria are emulsified with a loop in a drop of saline, and a small amount of antiserum added by means of the loop

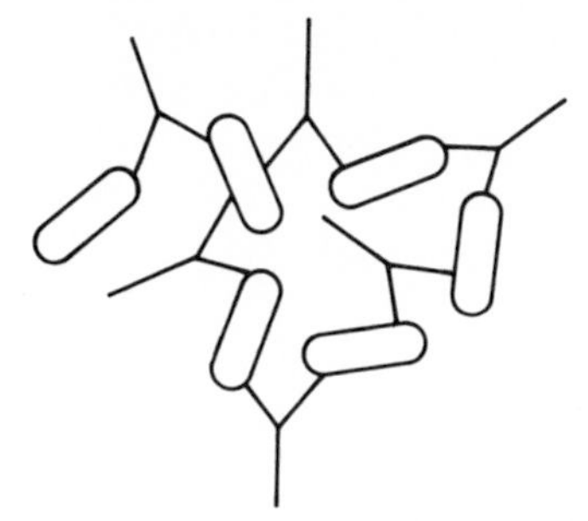

Fig. 62 The principle of cell agglutination by antibody molecules. The latter are Y-shaped, with combining sites for antigen binding on both the short arms of the Y, so that they can form cross-links between particles and hence will cause agglutination.

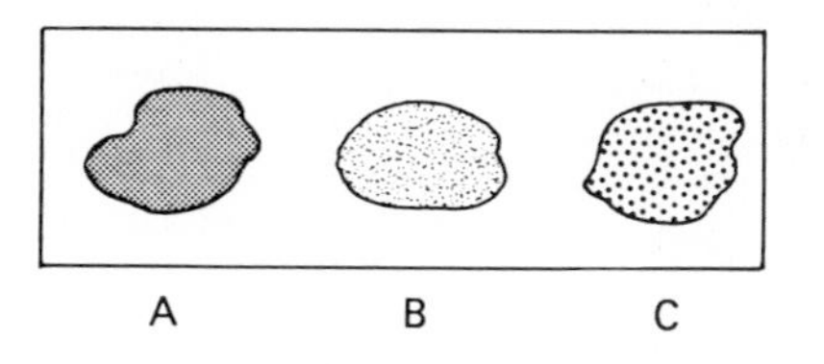

Fig. 63 Slide agglutination. Agglutination of bacterial or other cells on a glass slide is macroscopically visible. A control, unagglutinated suspension at (A) is uniformly turbid. A moderately agglutinated suspension shows barely visible particles as at (B). A strongly agglutinated suspension forms large aggregates with clear zones between (C). Sometimes the aggregates collect at the edge of the wetted area, leaving a clear centre.

and mixed in. The slide is tilted around for half a minute or so to allow any reaction to take place. Agglutination will be obvious as the initially smooth, milky suspension rapidly becomes visibly particulate, compared with a control preparation emulsified in saline (Fig. 63). Strong reactions produce larger aggregates. A problem arises when the organism will not produce a smooth suspension to start with, and the organism is then considered 'untypable'.

Agglutination tests may also be done in tubes, the so-called *Widal test*, in which dilutions of antiserum are prepared and mixed with a suspension of bacteria. After standing overnight, agglutinated bacteria are seen to have settled to the bottom of the tube, and a titre of the antiserum can be determined, and may be confirmatory of the specificity of agglutination seen with typing sera.

Agglutination tests have been used classically for typing enterobacteria such as *E. coli* and *Salmonella* strains, and for both species there are complex schemes of typing involving flagellar (H), capsular (K) and somatic or cell-wall (O) antigens. The K antigens can be further subdivided into subgroups depending on their thermal stability, i.e. whether or not agglutinability with typing antibodies is destroyed by boiling or autoclaving the cells before the test is carried out.

The principle of agglutination is also employed in haemagglutination, passive agglutination and co-agglutination tests. In *haemagglutination*, microbial antigens may bind specifically to molecules present on the surfaces or red cells from man or

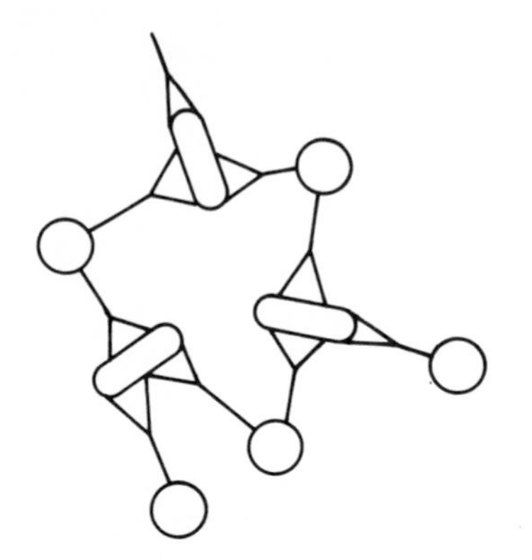

Fig. 64 Co-agglutination. Antibody molecules anchored to particles which do not directly take part in the immunological recognition mechanism (shown here as spheres) are co-agglutinated with the particles under test (in this case rod-shaped bacteria) due to the formation of linkages by specific interaction of antibody and antigen on the bacteria. The antibody carrier particles may conveniently be staphylococci, which bears the antibody-binding protein A, able to link specifically to the F_c portion of the antibody molecule (the stem of the γ which is not involved in antigen binding). Alternatively, latex particles may be used as antibody carriers.

other species, leading to cross-linking of the cells and a visible clumping reaction on slides or in tubes or wells in trays. Examples of haemagglutinating structures in microorganisms include fimbrial haemagglutinins of Gram-negative bacteria – many fimbriae, although not all, show haemagglutinating activity for erythrocytes of certain species – and the haemagglutinin molecules which occur on the surfaces of many viruses, notably influenza virus. In the case of bacteria, both the whole organism, at concentrations in the range of 10^7 – 10^9, and isolated fimbriae can show activity.

In *passive agglutination* techniques, either red cells or other particles such as latex beads of similar size are 'coated' with soluble antigen of microbial origin, and become the target for agglutination by specific antibody. *Co-agglutination* is a somewhat similar principle, but employs the reverse situation where antibodies are bound to particles, in this case to cells of *Staphylococcus aureus* which have a binding receptor protein (protein A) for the domain of the immunoglobulin molecule which is not involved in antigen recognition. The antigen-binding domains are then free to interact with specific antigens, and if such antigens are present, either as bacterial cells or as soluble antigen (provided they are large, multivalent complexes), the antibody–antigen reaction will take place, resulting in agglutination of the staphylococcal cells (co-agglutination if the antigen consists of bacterial cells) (Fig. 64).

PRECIPITATION IN GEL

Formation of antigen–antibody precipitates in agar or agarose gel has given rise to some powerful analytical techniques, not often used to their full potential. The basis for the method is that if antigen and antibody solutions diffuse into one another, a precipitate will form, as in mixed solutions, at the point of optimal proportions. As with agglutination tests, the phenomenon is based on the multivalency of the antibody molecules which enables them to form an insoluble complex aggregate with antigen.

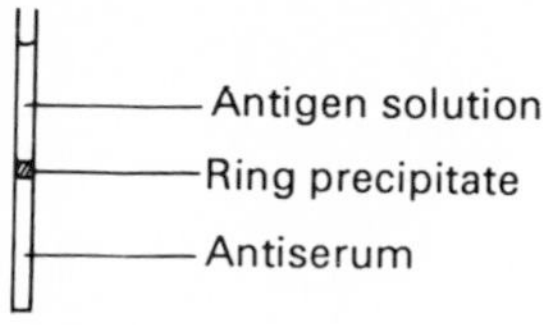

Antigen solution
Ring precipitate
Antiserum

Fig. 65 The capillary ring test. Antigen solution is layered carefully over the antiserum in a capillary tube, and an immunoprecipitate forms at the interface where it may be visible due to its increased turbidity.

This fact was first exploited in the *capillary ring precipitin test* (Fig. 65), in which liquid solutions are allowed to diffuse into one another after overlaying. Mixing of the two is retarded by the confined space in the capillary tube which reduces mixing by convection. By performing the reaction between wells in a layer of gel, however, there are several major advantages: the system is more stable; precipitates are easier to see, and there is scope for clearly separated multiple precipitates to form; and by juxtaposition of additional wells of either antigen or antibody, identical, non-identical or partially related reactions undergone by different preparations can be discerned (Fig. 66).

Methodology for precipitation in gel has been largely standardized by the use of barbitone buffer at pH 8.6 (barbitone sodium 8.76 g/litre, barbituric acid 1.38 g/litre; this can be made up as a stock solution at × 5 strength and diluted for use). Barbituric acid is slow to dissolve, and dissolves more easily in the alkaline solution of the sodium salt. Commonly, sodium azide (0.1 per cent w/v in working strength buffer) is added as a preservative, but it should be remembered that this is a strong oxidizing agent and may form azides of other metals if habitually poured down the same sink with, for example, copper plumbing; an explosive deposit is said to accumulate. Certainly antigen preparations, and especially serum, form a rich medium for growth of microorganisms if prolonged incubation is needed for reactions to develop, and a preservative of some kind is essential. The use of barbitone buffers at pH 8.6 is especially useful for the electrophoretic methods described below, since at this pH most antibody molecules carry little net charge and will not move rapidly in an electric field.

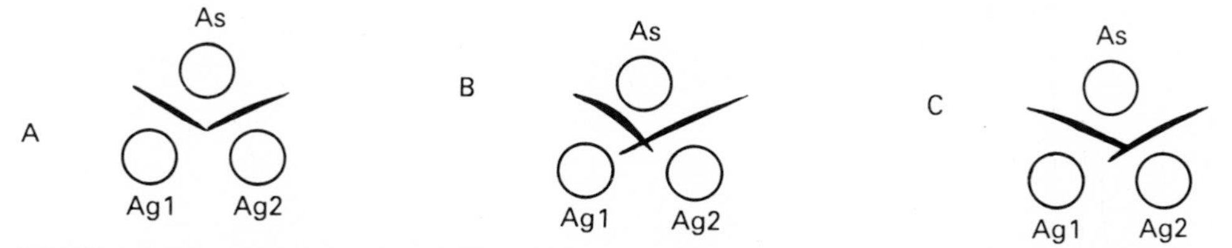

Fig. 66 Common reactions in the Ouchterlony double diffusion procedure, in which antibody (As – i.e. antiserum) diffuses through gel to meet antigen (Ag), with the potential to form a precipitate if the reagents are able to react. In A, antigens 1 and 2 are identical and form a reaction of identity in which one precipitate merges evenly into the other. In B, the two antigens responsible for forming the two precipitate lines are quite different, and there is no merging of lines to indicate a relationship. There must of course be antibodies to both antigens present in the antiserum. In C, the two precipitate lines merge partially, showing that there are antigens present in both preparations which share some but not all antigenic determinants (epitopes).

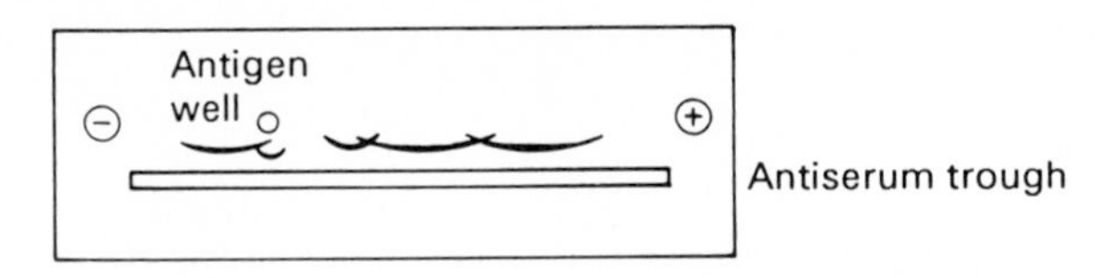

Fig. 67 Classical immunoelectrophoresis. Antigens applied to an agar or agarose gel in solution are electrophoresed from the sample well through the gel, until a marker dye (bromophenol blue added to the sample) is at the anodic (+) end of the slide. After electrophoresis, the antiserum trough is cut and filled, and antibody and antigen allowed to diffuse together to form precipitates as shown.

The format of test may vary: for *double diffusion*, agar in petri dishes was often used in the past, with wells of up to 1-cm diameter in agar 3–5-mm deep. Miniaturization is feasible, however, and leads to economy in use of reagents: wells of 3–5-mm diameter, 2–5-mm apart can be punched in a layer, on a microscope slide, of agar or agarose 1-mm thick. The closer the wells, the more sensitive the method for weak antigen or antibody preparations. Wells can be cut with thin-walled tube and the plugs removed by suction or by lifting out with a needle.

Several other techniques have been developed from the simple double diffusion in gel method. By combining electrophoreses of antigen in gel, with reaction of antibody either placed in troughs cut into the gel (Fig. 67), or incorporated into the gel during a second dimension of electrophoresis, the power of the technique for analysis of complex mixtures of antigens is greatly increased. Such mixtures are commonly obtained from microorganisms, and the use of these techniques allows their analysis, or the monitoring of perhaps a single component in complex mixtures, without the need for fractionation of the mixture and purification of the component concerned. Again, methods have been developed for comparison of the composition of different preparations. In addition, and again much under-exploited as blotting techniques (see below) have become fashionable, these gel methods have the advantage that very large

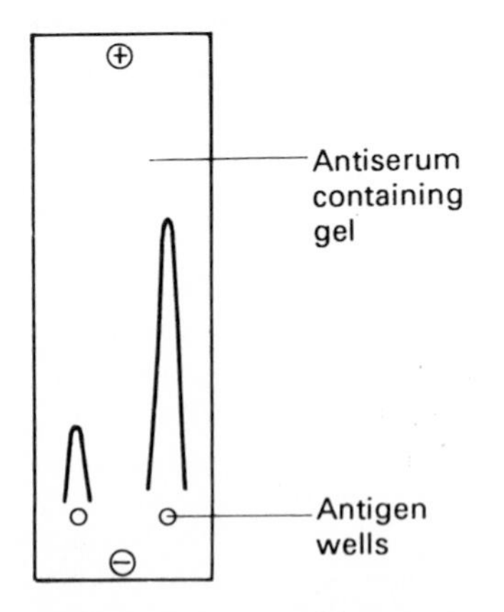

Fig. 68 A simple 'rocket' technique. Antigen is electrophoresed into a gel containing antiserum, towards the anode. Precipitates form in the gel and the 'height' from the well to the tip is proportional to the antigen concentration.

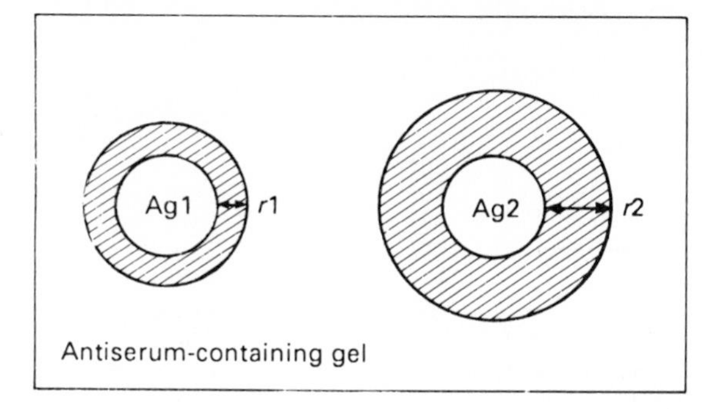

Fig. 69 Single radial immunodiffusion. Antigen diffuses out of wells cut in antiserum-containing gels, and because immunoprecipitates are soluble in excess antigen (an important aspect of all gel precipitation reactions), the precipitate forms a moving boundary as antigen diffuses from the well. The equilibrium position of the precipitate, indicated by the radii *r*1 and *r*2, obtained when the antigen is exhausted, is dependent upon the original antigen concentration (see text).

complexes of antigen, or even whole fragments of particulate antigens such as fimbriae or membrane vesicles with their constituent components, will enter and migrate through agar or agarose gels in their native state, whereas blotting techniques generally rely on denaturation of antigens prior to polyacrylamide gel electrophoresis.

Gel methods are also very useful for quantitation of antigen. This may be either by so-called *rocket immunoelectrophoresis*, in which the height of a peak of immunoprecipitate is proportional to the antigen concentration (Fig. 68), or by *radial immunodiffusion*, a much-used method in which antigen is placed in a well and allowed to diffuse out radially into a gel containing antibody (Fig. 69). The square of the diameter of the zone of precipitation formed under standard conditions is proportional to the antigen concentration.

An important aspect of all the electrophoretic methods used in gel-precipitation techniques is the nature of the gel matrix. Unpurified agar has a strong net negative charge, due to the presence of carboxyl and particularly sulphate groups, whereas agarose preparations have lesser proportions of these, and while generally less acidic, different preparations may vary. This very much affects their properties in electrophoresis work. Highly charged gels have a high rate of electro-endosmosis, i.e. flow of liquid through the gel due to interaction of the (mobile) charged particles in the buffer with (static) charged residues on the gel matrix. In low-endosmosis gels, at pH 8.6, there is little flow and antibody molecules which have approximately neutral charge will remain stationary: the ideal for rocket or two-dimensional analysis, for which crude agar preparations are not suitable as the gel matrix. At this pH the vast majority of proteins of microbial origin have a net negative charge, and so will move towards the anode and into the antibody-containing gel. However, slightly more charged gels will be advantageous in some circumstances since the flow of liquid may carry antibody molecules in the opposite direction to the electrophoretic motion of the antigen. This is exploited in the technique of *counter-immunoelectrophoresis* (CIE) (Fig. 70), whereby the sensitivity of simple diffusion in gel is enhanced and the reaction accelerated by the electrophoretic effect. Different agarose preparations of controlled endosmotic properties are obtainable.

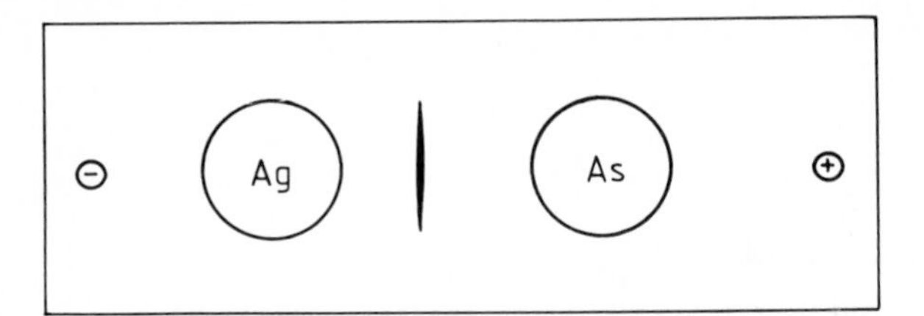

Fig. 70 Counter-immunoelectrophoresis (CIE). Antigen and antiserum are encouraged to react with great speed and efficiency by adding an electrophoretic element to a reaction which might also occur by simple diffusion. Antigen migrates electrophoretically towards antibody contained in the As well; the technique relies on the net negative charge on most antigens at pH 8.6, whereas antibody molecules are generally electroneutral at this pH and are carried passively by the electro-endosmotic flow of buffer towards the antigen well.

ELISA ASSAYS

A whole battery of tests based on the ELISA (*E*nzyme-*L*inked-*I*mmuno*S*orbent *A*ssay) have been developed in recent years. This is a very powerful technique due to both its high sensitivity and its quantitative nature. The revolution in serology which it has spawned derives from the use of a solid phase as the site of the reactions involved, and exploitation of enzyme reactions with chromogenic substrates which yield coloured reaction products, visible as well as measurable by spectrophotometry at very low concentrations. Some basic formats for the test are shown in Fig. 71, although many other variations are possible. The principle is that either antigens or antibodies will bind spontaneously and non-specifically to new plastic surfaces, usually of polystyrene, which have not been exposed to biological substances or surfactants. The binding is probably mainly through hydrophobic interaction, and once the component of the test which is intended to bind is in place, further binding of any other reagents used in the test is prevented by saturating the binding capacity of the surface with an inert protein or mixture of proteins, or by exposure to a surfactant which will interfere with hydrophobic interaction.

Thus, the first step in an ELISA assay is binding of the first reagent, commonly antigen, to the plastic surface. The antigen may be whole bacteria, or particulate or soluble antigens derived from them (or, of course, in a wider context, any other antigenic protein). Commonly binding is allowed to take place at an alkaline pH of 9–10, buffered by, for example, sodium bicarbonate (0.05 M, pH 9.6) or ethanolamine, in flat-bottomed polystyrene microtitre plates. Binding may take, for example, 1 h at 37°C, or overnight incubation at room temperature or 4°C; the latter, however, may allow binding which is unstable at room temperature and is therefore difficult to sustain during subsequent manipulations and may be variably lost at higher temperatures – a consideration which applies at all subsequent stages. After initial binding, the capacity of the surface to bind any further protein is blocked by incubation with a 'neutral' protein such as albumin (1–10 per cent w/v in PBS containing 0.05 per cent v/v Tween 20 – a surfactant). Many authors refer to blocking of 'binding sites' as though there were specific locations for binding of proteins, implying a degree of specificity in the process which probably does not exist: 'binding capacity' is perhaps a better term. Albumin, usually bovine, is quite an expensive

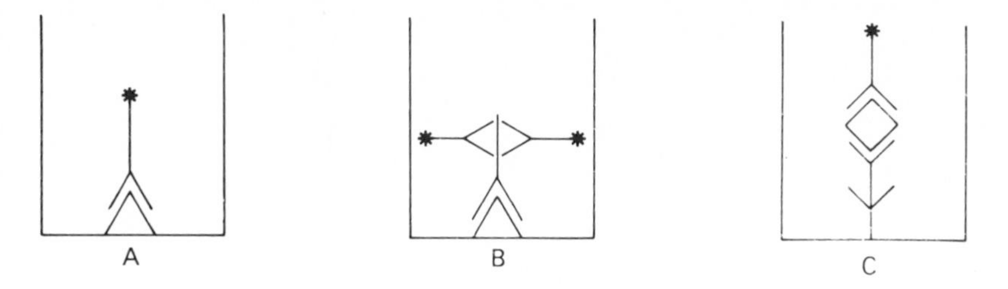

Fig. 71 Different configurations for ELISA testing. At A, the most simple arrangement, antigen bound to the solid phase reacts with antibody which is conjugated to an enzyme capable of performing a chromogenic reaction, as shown by *. At B, a double antibody system is used, in which the first antibody reactive with antigen is followed by a second antibody (raised in a different species against species-specific determinants on the first antibody) which is enzyme-conjugated. The potential for amplification by a second antibody is clear. At C, an example of the very complex configurations possible with ELISA tests is shown; the first antibody is a 'capture' reagent bound to the solid phase, which binds antibody, perhaps of a specific class, from the serum under test. The antigen is then added (square particle), and if bound, will be detected by a specific antibody, which in this case is conjugated to a colour-generating enzyme.

protein even if not fully purified ('Cohn fraction V' from ammonium sulphate precipitated serum proteins is commonly used), and as it is used at high concentration in ELISA work, cheaper substitutes such as skimmed milk powder (e.g. Marvel) at 2.5 per cent w/v are often used and are said to be equally effective.

Subsequent binding reactions in ELISA assays should all be specific reactions between complementary structures involving proteins, with rigorous washing between one or more successive steps to eliminate any traces of the protein which could lead to spurious signals. Incubations for binding are commonly for 30–60 min at 37°C. Intermediate washes with PBS containing 0.05 per cent Tween 20 and/or 0.1 per cent BSA are done, for example five times. It is convenient to use a wash bottle to fill the wells in the microtitre plate, and to shake the plate upside down over the sink to remove the wash fluid; any remaining drops can be expelled from the wells by inverting the plate with a firm impact onto clean paper towels on the benchtop. If large numbers of plates are being processed routinely, these procedures can be done automatically by machines with dispensing and suction manifolds in the 96-well format, which can be inserted into the wells.

After all protein binding steps are completed, enzyme substrates are added and colour allowed to develop until optimal – usually for minutes rather than hours. Enzymes used commonly in conjugates are horseradish peroxidase, alkaline phosphatase and urease. The conjugates and their substrates can be obtained commercially from a variety of sources. Colour development can be stopped by addition of strong acid or alkali, depending on the system, so that quantitative comparisons can be made between wells which have all been incubated for the same length of time. Results can be read quantitatively and objectively by a plate reader, a multi-channel colorimeter which can operate at various wavelengths depending on the substrate used: a highly efficient instrument which can take readings for a whole plate, and print out 96 results on, for example, thermal paper strip, in one or two minutes. The accuracy of readings is however limited and should be assessed for each system used.

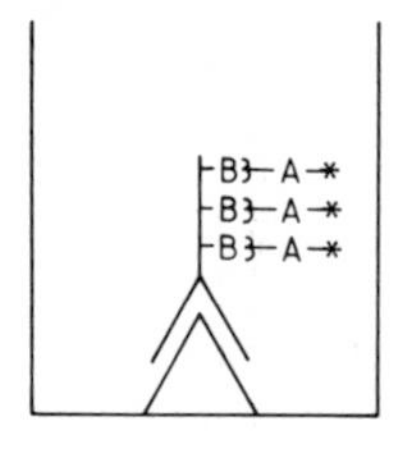

Fig. 72 The principle of biotin labelling of antibody molecules. Biotin (B) is a low-molecular-weight ligand which can be bound at multiple sites on an antibody molecule without affecting its antigen-binding capacity. After use in a conventional ELISA test, the biotinylated antibody can be detected by avidin (A), which binds very specifically and strongly to biotin, and in turn bears the colour-generating enzyme.

ELISA assays are often most sensitive and versatile when multiple layers of reagents are built up. The sensitivity of an 'indirect' format results from the amplification obtained by reacting a 'first' antibody with immobilized antigen, followed by a 'second' antibody directed against antigenic components of the first antibody. Because more than one molecule of second antibody, each with conjugated enzyme, can bind to each molecule of first antibody, amplification of the enzyme-mediated signal results. An alternative final step in the reaction sequence is the exploitation of the avidin–biotin interaction, a highly specific, high-affinity glycoprotein–ligand interaction which offers a superior alternative to the second antibody conjugate (Fig. 72). The second antibody is linked to a number of residues of biotin, and this is then bound very effectively, with further amplification, by an avidin–enzyme conjugate.

IMMUNOBLOTTING

Immunoblotting is an extension in concept of the gel electrophoretic methods of analysis described above for analysis of complex mixtures of antigens, and the monitoring of antibody reactions with individual components in such mixtures. The best obtainable electrophoretic separations of complex protein mixtures have been found to depend on molecular weight criteria, rather than charge which is essentially the criterion for separation in agarose gels as described above for immunoelectrophoresis. Agarose gels in the concentration ranges commonly used in immunoprecipitation in gel are essentially freely permeable for most proteins, with no significant retardation in electrophoretic migration caused by molecular size and gel-filtration effects. In systems which do depend on the latter, historically in starch gel electrophoresis and latterly in polyacrylamide gel electrophoresis, the use of antibodies as probes for specific proteins derived from complex mixtures is not possible: the antibody molecules are too big (even IgG has a molecular weight of about 170 000) to enter the gel matrix and react with antigens. Therefore to combine the advantages of the exquisite resolution of protein mixtures obtainable with sodium dodecyl sulphate polyacrylamide gel electrophoresis (SDS–PAGE) and the specificity of antibodies as probes for antigenic components, immunoblotting is the ideal solution. If proteins are transferred, after separation and in the same banding pattern as in the gel, to nitrocellulose paper (or membrane – NCM) to which they will adsorb by electrostatic and hydrophobic interaction as with the solid phase in an ELISA assay, antibodies in free solution can interact unhindered with the surface-located antigens. This is an immensely powerful analytical tool and has become one of the most used methods for antigenic analysis of microorganisms.

The method was initially based on the initial electrophoretic separation followed literally by 'blotting' the separated proteins from the separation gel to NCM: a sandwich of gel and paper between filter papers and paper towels to absorb moisture was pressed together for several hours, and moderately efficient transfer of proteins resulted. The method was much improved however by the introduction of *electroblotting* – electrophoresis was used to effect the transfer from one medium to the other, perpendicularly to the plane of the first separation (Fig. 73).

Details of the methodology used for electroblotting are too complex to include here, but it is appropriate to give some idea of the range of equipment and technical expertise needed: often the introduction of powerful modern techniques is deferred because the technology seems dauntingly complex to the newcomer to the field. First, the chemicals and materials for SDS–PAGE are basically simple, although for making up polyacrylamide gels it is important that the ingredients are of high quality: electrophoresis grade, from a reputable supplier. Descriptions of the methods are easy to find (see Further Reading list). The equipment can be quite simple: miniaturization is an ongoing process, and small SDS–PAGE gel tanks can be bought for £200–300. A stable power supply is needed for electrophoresis, and for use with mini-tanks simple devices are available with limited voltage outputs, again for about £200. Blotting can be done with very cheap, improvised apparatus: the gel–paper sandwich can be clamped between 'Scotchbrite' pads held in place by perforated (large holes) rigid plastic sheets, and immersed in a tank with electrodes of platinum wire no more than about 10 cm apart. The tank should be as small as practicable to limit the cross-sectional area of electrolyte between the electrodes, and hence to limit the conductivity and the current carried with a reasonable transfer voltage. Because the current will be

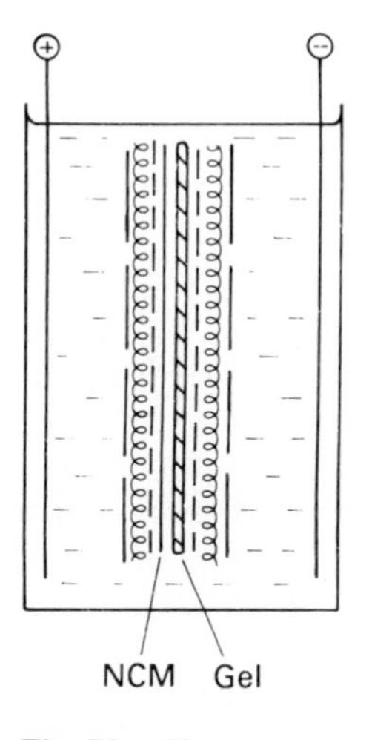

Fig. 73 The principle of electroblotting. Antigens (or other macromolecules) separated by gel electrophoresis are transferred from the separating gel to a nitrocellulose membrane (NCM), by electrophoresis at right angles out of the gel towards the anode, and into the NCM to which they bind.

higher for this reason than for most laboratory electrophoretic applications, either a special high-output power pack or, for low-budget work, an ordinary car battery charger can be used: a voltage gradient of about 0.25 V mm^{-1} (12 V over 5 cm) will be quite adequate for effective transfer overnight. Transfer is performed in a mixture of methanol and tris-HCl buffer; care should be taken not to allow the temperature to rise during transfer, otherwise an accelerating process of methanol evaporation, increased conductivity as ions become more concentrated, and increased electrical heating may follow, with catastrophic results!

The detection of antigens on *nitrocellulose membrane* (NCM) follows basically the same principles as the ELISA technique: antigens are adsorbed to a solid phase, and in exactly the same way, additional adsorbing capacity is blocked with inert protein or surfactant, and a series of washes and reactions with antibody and enzyme conjugate can be used to label and locate the antigenic constituents. Handling of the nitrocellulose before blotting must be done in scrupulously clean conditions, since traces of 'foreign' antigens are easily adsorbed and will make the final result appear dirty. Again, indirect techniques with a second antibody conjugate are commonly used. The reagents, particularly enzyme-conjugated second antibodies, used in such work are essentially similar to those used for ELISA, and are quite expensive although they go a long way. NCM itself is also a major expense. Differences in substrate systems will be needed, however, to give insoluble reaction products which will adsorb to the membrane, rather than soluble products which will be washed away!

Nitrocellulose membrane is very useful also as a matrix for other solid-phase assays. Colony replicas are mentioned in Chapter 4, and can be used for immunological detection of antigens expressed by particular colonies of organisms, usually through the use of monoclonal antibodies. Another technique which is commonly used for multiple tests with a shared sequence of processing manipulations is the *dot-blot*, in which small volumes of a series of antigen preparations can be 'dotted' onto NCM and allowed to adsorb, the NCM then being processed as above to reveal the presence of a specific antigen. Recording can be by photography (most easily processed mentally as an analogue image of the protein separation), or by scanning with a laser densitometer to give a series of peaks. A trace from the latter can be processed digitally.

Genetic manipulation

The basic procedures involved in genetic manipulation are preparation of DNA from whatever source is involved, and its incorporation (recombination) into a replicating vehicle or vector so that multiple identical ('clonal' – hence the term 'gene cloning') copies of each sequence involved can be produced following introduction of the recombinant vector into a suitable 'host' for its replication to occur. In the following examples, it is assumed that the DNA to be manipulated is of microbial origin (likely to be the major interest of intending microbiologists!), but the principles are similar whatever the source.

SETTING UP THE LABORATORY FOR GENE CLONING

One of the most important principles of work with recombinant DNA is that the conditions, equipment and reagents used should all be of the highest possible standards of cleanliness, purity and quality. Due to the power of the techniques involved to

Table 11 Examples of standard buffers commonly used in molecular biology.

Abbreviation	*Composition*	*Uses*
TE	10 mM Tris HCl, 1 mM EDTA pH 7.4–8	Handling and storage of DNA. EDTA chelates divalent cations to inactivate nucleases
TNE or STE	As above plus 100 mM NaCl	Higher ionic strength applications
TBE	89 mM Tris-borate, 89 mM boric acid, 2 mM EDTA	Standard buffer for agarose electrophoresis of DNA
SM	100 mM NaCl, 12 mM $MgSO_4$, 20 mM Tris HCl, 2% gelatin	Phage storage and dilution

amplify any DNA present in the system, whether it is the intended target or not, and the sensitivity of many of the enzymes and procedures used to spoilage or failure due to contamination or incorrect conditions, attention to conditions and procedures cannot be over-emphasized. In addition, of course, rigid control for safety reasons is necessary, as outlined in Chapter 6.

All equipment and plastic and glassware must be both clean in the ordinary chemical sense and free of nucleases – enzymes which destroy nucleic acids. Nucleases are ubiquitous in the environment, since they are produced by all living cells. Some nucleases are very stable molecules. It is therefore necessary to prepare all equipment and materials with great care; disposable plastics should be used wherever possible, and discarded after one use. Eppendorf tubes and micropipette tips for handling DNA and performing enzyme digestions and reactions, normally made from heat-stable polypropylene, should be autoclaved before use. Sterile disposables should be handled in such a way as to keep them sterile, and therefore uncontaminated. A source of ultra-pure water is necessary, normally either distilled and de-ionized or processed to molecular biology grade in specially designed equipment by a combination of ion exchange and ultrafiltration. It should also be autoclaved before use.

Several buffers and salt solutions of standard formulation are used in molecular biological manipulations, and are often prepared as concentrated stock solutions, autoclaved for storage, and handled for dilution or mixing for use essentially as sterile solutions so that they remain uncontaminated. Standard abbreviations for these solutions are also used commonly. Examples are given in Table 11. Reagents, for example restriction and other enzymes, are commonly supplied in Eppendorf tubes as solutions in 50 per cent glycerol for storage at −20°C, and can be stored in insulated freezer boxes so that exposure to higher temperatures when sampling the contents is minimal.

The basic equipment for genetic manipulation must include facilities for autoclaving and media preparation, a static incubator for petri dishes, shaker incubation for liquid cultures, water baths and heating block for short-term incubations, a set of micropipettes and a vortex mixer, a microfuge and a benchtop centrifuge, a microscope, and gel electrophoresis equipment and an ultraviolet illuminator for viewing agarose gels in which DNA has been analysed (eye and skin protection against ultraviolet irradiation is essential). A polaroid camera for recording the results is also essential. Access to an ultracentrifuge for purification of DNA on a reasonable scale will also be necessary from time to time.

CULTURES

Cultures grown for genetic manipulation are basically no different from any others. Solid media are often used with antibiotics, and it should be remembered that these are not necessarily stable once formulated into media: penicillin, for example, is notoriously liable to become inactive after just a few days at ambient or incubator temperatures. Liquid cultures may be required in the logarithmic growth phase – usually 'late log' since yields will be higher at this time; great care is needed if cultures are not to go 'over the top' beyond the end of the log phase. Maintenance and storage of cultures is described in Chapter 13. In addition, care is needed in verifying that particular mutants or host strains are indeed what they are supposed to be. After verification, for example by checking auxotrophy on minimal media with and without the relevant nutrient, or for enzymic markers by the use of appropriate chromogenic substrates, or for antibiotic resistance by checking for growth in the presence of the agent, cultures of specified strains should not be subcultured routinely for use, but inoculated afresh from verified stocks when needed, in case repeated subculture should lead to accidental selection of atypical mutants or revertants.

Many standard strains of host bacteria, as well as standard cloning vectors, can now be obtained commercially. Bacterial strains widely used in standard procedures may also be obtainable from culture collections. In addition it is common (though not universal!) practice for authors to supply (within reason) novel reagents or strains to fellow researchers once they have published detailed protocols for new procedures.

OBTAINING DNA

Genomic DNA can be obtained with varying difficulty from microorganisms depending on their structure. Commonly, cells are simply lysed to release DNA, which must be protected from enzymic degradation from then on. Nucleases commonly require the presence of divalent cations for their activity, and therefore EDTA is routinely included in all buffers to chelate any divalent cations present and inactivate nucleases. In addition, the main contaminant of significance in any DNA preparation is protein, including not only nucleases but also proteins capable of binding DNA in various ways which may interfere with its manipulation. Two procedures are commonly used for the removal of protein. First, it can be denatured by phenol, which added part for part in liquid form (molten phenol is saturated with a suitable aqueous buffer and cooled) to aqueous solutions of DNA causes denaturation of proteins which collect at the interface of the resulting two-phase system after emulsification and separation and centrifugation into two phases; DNA remains in the aqueous phase, and residual phenol can be removed by extraction with chloroform. Secondly, proteins can be completely digested by powerful proteases such as protease K, obtained from a streptomycete, which will completely digest virtually all proteins and thus inactivate any nucleases present as well as removing any bound or potential DNA-binding proteins present in the preparation. Similar steps to remove protein are generally necessary in the processing of DNA by enzymic reactions, when the DNA product has to be freed of the enzyme before the next step in a protocol.

DNA release from bacteria depends on lysis of the cell. Traditionally, this has been done by the action of detergents, following if necessary the digestion of cell-wall polymers with appropriate enzymes. Some Gram-negative bacteria may easily be lysed by detergent alone in the presence of EDTA (the latter may destabilize outer-membrane structure by removing Mg^{2+} ions as well as inactivating nucleases). Often, however, protocols specify lysozyme digestion even for Gram-negatives. For Gram-positives, lysozyme digestion may be essential, for example for *Bacillus* spp., or other enzymes may be needed to break down the cell wall of, for example *Staphylococcus*, which is resistant to the action of lysozyme, and requires the action of the bacterial enzyme lysostaphin. To make protoplasts of fungal cells such as yeasts, digestion with beta-glucanases and/or other, poorly characterized enzymes including zymolase and lyticase, with the additional action of reducing agents (2-mercaptoethanol of dithiothreitol) may be required before detergent-mediated lysis can be achieved.

Following release of DNA from the cell and removal of protein by whatever means, it must be concentrated and stored in stable form. Precipitation in the presence of sodium ions, in the form of, for example, 0.25 M final concentration of sodium acetate, by two volumes of ice-cold ethanol at −20°C is the standard procedure. After about 1 h, the DNA can be pelleted in a microfuge (10 min), drained and finally dried in a vacuum desiccator. It can be stored dry, or will be stable for some days at 4°C in solution in TE: if it is to be digested subsequently, sodium chloride should be absent from the solution so that the desired salt level for enzyme action can be achieved subsequently.

PREPARATION OF PLASMID DNA

Plasmid DNA can be prepared relatively easily, especially in the case of small plasmids (less than about 25 kb), often of high copy number. Because the bacterial chromosome has a very high molecular weight (in *E. coli* approximately 2×10^9 daltons, or 3×10^6 base pairs), it is effectively impossible to handle in solution without physical breakage into smaller fragments, caused by shearing forces as the liquid is moved during handling. On the other hand, plasmid DNA can remain undamaged in solution, because of its much smaller size (usually approximately 2–100 kbp), and if neither nicked (broken in one strand only, allowing rotation about the single remaining covalent bond and hence uncoiling) nor broken, it will retain a supercoiled, covalently closed circular (CCC) conformation. This form differs physically from linear chromosomal fragments in several important respects. It is less easily denatured by alkaline conditions, and hence will remain soluble after exposure to NaOH (and SDS) at pH 12 followed by neutralization; linear DNA and protein tend to be precipitated under these conditions and can be removed by centrifugation, leaving plasmid DNA and a small amount of protein in solution – a so-called 'cleared lysate', and the basis of many rapid plasmid preparation methods, notably that introduced by Birnboim and Doly in 1979. Secondly, supercoiled CCC DNA is less able to take up intercalating dyes such as ethidium bromide, which insert into the double helix and alter the structure of the molecule, than is the linear form; it therefore has a higher density than linear DNA in the presence of the dye, and the two can be separated by buoyant density centrifugation. This is usually performed by ultracentrifugation in a self-forming caesium chloride gradient. After centrifugation, either for at least 18 h in a conventional ultracentrifuge at about 40 000 r.p.m., 150 000 g or for 3–4 h using a small-scale benchtop ultracentrifuge, caesium chloride can be removed from the recovered plasmid fraction by dialysis. An increasingly used alternative to density gradient centrifugation is chromatographic separation of circular and linear DNA.

PREPARATION OF PHAGE DNA

For this purpose, phage particles have to be produced in large amounts. The most commonly used phages for genetic manipulation are strains of bacteriophage lambda. The strains have to be propagated and isolated by plaque picking analogously to bacterial strains. Plaques are produced on lawns of 'plating' bacteria. These are prepared from stocks of suitable strains of *E. coli* grown overnight in a rich liquid medium such as LB broth, supplemented with 0.2 per cent maltose to induce the lambda-receptor outer-membrane protein, pelleted and suspended to half the original culture volume in 0.01 M $MgSO_4$, and 0.1 ml added to 3 ml LB top agar (0.7 instead of 1.5 per cent w/v agar) and poured onto the top of ordinary LB agar plates. After setting, the plates can be streaked in the normal way with phage stocks, or plated by dilution methods as for a viable count of bacteria, incubated overnight and isolated plaques picked for further use. Phage lambda is commonly stored and manipulated in SM medium, a NaCl–$MgSO_4$ buffer, containing 5.8 g NaCl, 2 g $MgSO_4 \cdot 7H_2O$ and 0.1 g gelatin per litre of 0.05 M tris-HCl, pH 7.5. Plaques may be picked by suction removal with a pasteur pipette, and phages released by soaking in 1 ml SM medium containing a trace of chloroform. Small-scale stocks can then be produced by adsorption of this initial preparation to stationary-phase cells prepared as above, and either preparing fresh plates from them or inoculating a suitable volume of liquid medium; incubation for 8–12 h should lead to lysis of the entire culture and release of the phage particles. There is some skill involved in judging the correct ratio of phage to bacteria, even more important when phage are grown on a larger scale, if the yield and concentration of phage is to be optimal. Detailed protocols are given by Maniatis *et al.* (see Further Reading). Phage particles are purified, usually by density gradient centrifugation. DNA is then prepared, following disintegration of the phage by EDTA treatment, and protease, SDS and phenol treatment, broadly in line with that described for plasmid DNA.

ANALYSIS OF DNA

DNA is routinely analysed by agarose electrophoresis. Agarose gels (molecular biology grade agarose) are made up usually to between 0.7 and 1 per cent in tris-borate-EDTA (TBE) buffer, and poured several millimetres thick. Wells formed by carefully machined combs do not penetrate right through to the base of the gel holder, since leakage of DNA between plastic or glass and the gel could occur. Gels are usually run horizontally, submerged in a shallow layer of TBE to prevent desiccation during the run and to prevent localized differences of temperature. Minigels, about 8–10-cm long, can be run in 1–2 h. After electrophoresis, results are monitored by ultraviolet illumination of DNA in the gel, stained with ethidium bromide, which may be either incorporated in the gel and buffer during the run at low concentration (0.5 $\mu g\, ml^{-1}$), or the gel may be soaked in the same concentration of stain after the run for 0.5–1 h. Ultraviolet illumination may be either from above, or more commonly with a transilluminator with a special purple glass filter able to transmit ultraviolet while retaining most of the visible light. Photography with a red filter then allows recording of the fluorescent image. Care should be taken not to get burned by the powerful ultraviolet illumination, nor to come into contact with ethidium bromide which is a potential carcinogen.

ENZYME DIGESTIONS

When DNA from vector and genomic sources for cloning experiments has been obtained, enzyme treatments, ligations, etc. are normally performed using the basic tools already described. Eppendorf tubes are usually the reaction vessels. Reagents (DNA, enzymes, concentrated buffers, water, etc.) are deposited on the side wall of the tube by micropipette. Reaction volumes are usually very small – often a total of just 10 μl for a restriction digest, for example. Pulse spinning in the microfuge, for just a second or two, will deposit the reagents together at the bottom of the tube. They can be briefly mixed by vortex or by flicking the tube, re-deposited, and the tube then incubated (e.g. floated in a water bath) for the reaction to take place.

TRANSFORMATION OF BACTERIA BY RECOMBINANT DNA

When recombinant DNA has been obtained by the digestions, ligations and other treatments necessary *in vitro*, it must be introduced into a replicating host cell, most commonly *E. coli* for many routine purposes (although not all foreign DNA will necessarily be stable or capable of expression in this organism and there may be better alternative bacterial species). Usually DNA is introduced into *E. coli* by transformation, in which the cells actively take up DNA under certain conditions. Conditions which favour DNA uptake tend to enhance the closeness of proximity between the cells and the DNA: presence of Ca^{2+}, for example, presumably by acting as a divalent linkage between the acidic cell surface and DNA, seems to enhance DNA uptake very effectively. High concentration of DNA is also advantageous. *Escherichia coli* is not, in fact, a particularly competent (i.e. highly transformable) species, hence the necessity for these procedures. Log-phase cells are particularly suitable for transformation. Normally a selective pressure (an antibiotic) is applied to transformants, so that parent cells are killed and only transformants carrying the recombinant plasmid expressing a resistance gene will survive.

A recently introduced alternative to transformation for introduction of DNA into cells is *electroporation*. This can be extremely efficient, and bacterial species which are normally resistant to transformation can be made effective targets. The basis of the technique is very simple – high-voltage pulses applied across a suspension of DNA and target organisms simply appears to cause physical impact between DNA and the cell surface, resulting in enhanced uptake of DNA. The equipment needed is purpose designed but is basically simple in concept.

NEWER TECHNOLOGIES

It should be clear from the above outline of the main procedures used in genetic manipulation that there is no real technical barrier to any competent microbiologist with some experience of modern protein and nucleic acid separation and purification technology in performing recombinant DNA work. Indeed there are always new techniques being developed, many of which will make it even easier to adopt these approaches to problems in biology. Many commercial suppliers are providing almost all the special reagents needed, including, for example, the bacteria competent for transformation as described above, supplied frozen and ready for use after thawing. Kits for enzyme digestions, DNA handling and sequencing, etc. are extremely

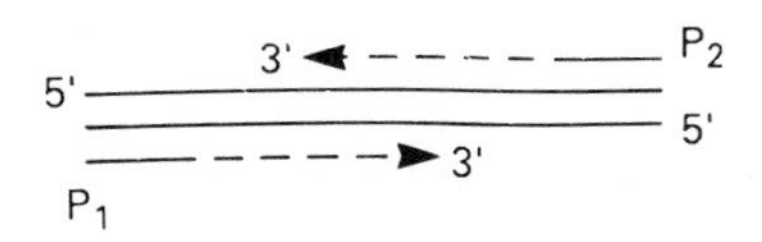

Fig. 74 The polymerase chain reaction. By utilizing primers for DNA replication which bind to opposite strands at a suitable distance apart, new copies of each strand are generated by DNA polymerase.

sophisticated and offer amazing short cuts to the laboratory with a large budget! There are, for example, several alternatives to the ultracentrifugation described above for the purification of plasmid and other DNA, based on affinity or ion-exchange or size-exclusion chromatography, often with remarkable miniaturization.

New developments in DNA amplification and separation have also emerged recently, which will also revolutionize some approaches to the problems which can be tackled by molecular biological approaches. The *polymerase chain reaction* (PCR) is an ingenious method for the amplification of specified sequences without going through the entire cloning protocol. By the use of single-stranded oligonucleotide primers (located on opposite strands) derived from separated, paired tracts of known sequences, intervening sequences of DNA can be synthesized using scarce copies of the desired target as template for DNA polymerase. By using heat-stable polymerase (Taq polymerase) obtained from the highly thermostable bacterium *Thermus aquaticus*, repeated cycles of synthesis, melting by heat treatment, annealing of new primers, and repeat synthesis, doubling numbers of the target sequence can be synthesized at each cycle in logarithmic (base 2) fashion (Fig. 74). The technique has great promise for use in sequencing, amplification of wild-type or mutagenized regions of genomic DNA from bacteria, detection and manipulation of target sequences in pathological and environmental specimens, and numerous other instances in DNA manipulation.

Another technical advance has occurred recently in the analysis of very large tracts of DNA. It is clear from the above discussions that DNA in solution is very unstable above a certain size. However, if it is released from organisms *in situ* in gels (for kinetic stability), and if the electric field for electrophoresis is specially pulsed in such ways that very large molecules of DNA can be induced to migrate by flexion through the gel pores, they can be analysed effectively in sizes essentially representing the entire chromosome, or very large fragments of it. The technique is currently in its infancy and it is technically difficult to obtain very clear results, but it appears likely that it will revolutionize, for example, mapping of genes on chromosomes. Newly discovered 'rare cutting' restriction endonucleases which recognize sequences of perhaps 8 bases and which therefore cut at random only every 4^8 (65 376) bases on average, provide tools for specific cleavage to yield large DNA fragments, and in combination with large-fragment analysis and Southern blotting, it should be possible to identify the locations of (i.e. to map) specific sequences on the scale of the entire microbial genome, a very powerful approach in terms of time and effort involved compared with classical gene mapping.

CHAPTER 13

Microorganisms commonly used in the laboratory

In this chapter, a selection of the microorganisms more commonly encountered and used experimentally will be introduced, with a little background knowledge of the genus and some brief indications of their value as experimental systems and the considerations involved in handling them. As in many branches of biology, the selection is driven by the practical needs and interests of the scientist, and hence there is some bias towards organisms of medical or public health importance, reflecting the interests of the author. The intention is not to give an exhaustive treatment to each example, but to highlight aspects of practical importance which will help the beginner to start work with an organism not previously encountered.

Bacterial classification and taxonomy

Before considering a list of major genera or species, a few comments on classification may put the problems of diversity and how we handle it into perspective. Bacteria are very diverse: their range of guanine plus cytosine content as a proportion of the total bases in the DNA, for example, is wider, at a range from 25 to 80 per cent, than is found in any other biological group (vertebrates, for example, are all clustered around 35–40 per cent). Modern molecular methods for characterization of phylogenetic markers useful in classification, such as ribosomal RNA sequences, are suggesting that many of the criteria used traditionally for the classification of bacteria have little relevance to their true pattern of evolutionary divergence. It is tempting therefore to be dismissive of the currently accepted and widely used systems of classification and nomenclature as being of little real biological significance, and unworthy of close study. Nevertheless, the system has been derived to fill a real need 'at the sharp end' – it is necessary for practical purposes to know the potential dangers, for example, of handling enteric Gram-negative organisms in the laboratory. Usually the empirically identified criteria of speciation are accepted precisely because the result is a system which works – if a Gram-negative bacillus isolated from the gut ferments lactose, we can be reasonably sure it is the relatively harmless (usually) *Escherichia coli* and not a highly pathogenic strain of *Shigella* or *Salmonella*, even though it may be so closely related to them in terms of phylogenetic origin as barely to justify its status as a separate genus. Classification may seem unimportant if we are working with a well-defined and thoroughly investigated organism, but it becomes an important issue when we find a new, fascinating organism with novel pathogenic or metabolic properties – what is it, and how does it relate to the current body of knowledge?

The enteric group

This is a large and diverse group of small, Gram-negative, rod-shaped aerobic and facultatively (i.e. optionally) anaerobic bacteria, of which the prototype *Escherichia coli* is the best known example. Not all the members of the group are actually associated with enteric (digestive tract) habitats however. Some are – including the very closely related *Salmonella* and *Shigella*, known with *Escherichia* as the coliforms; others including *Proteus*, *Enterobacter*, *Serratia* and *Klebsiella* are more normally found in soil and water, while other associated genera such as *Yersinia* and *Vibrio* are often pathogens associated with man or animals.

Escherichia coli

This most-used and best-known of experimental models in molecular biology merits a detailed introduction. In addition to the above characters of the group, it is distinguished by motility derived from peritrichous (i.e. multiple, uniformly distributed rather than polar) flagella, and ability to ferment lactose and possession of beta-galactosidase. The latter ability, or lack of it, is a distinguishing feature among the coliforms, and is the basis for use of MacConkey agar to identify them on isolation: lactose and neutral red are present in the medium, so that fermentation of lactose with acid production results in local accumulation of acidic products of fermentation and a resulting colour change to bright magenta around the colonies.

The organism is nutritionally undemanding, and will grow, albeit at reduced rates, in minimal media. It thus has a wide range of metabolic capabilities. It is robust – able to withstand a variety of environmental conditions, including survival in water, hence its transmission from animal to animal or man to man in natural waters; it is, however, killed rapidly at elevated temperatures, for example in a few minutes at 56°C. It grows rapidly under ideal conditions, with minimum doubling times of 15–20 min. On rich solid medium, it will produce colonies of 1–2-mm diameter after overnight growth. The colonies tend to be round, smooth and sharply defined, with a translucent pale grey colour. The organism has a sizeable genome, probably about average for bacteria, of about 2×10^6 base pairs (about 3×10^9 molecular weight) in a single circular chromosome which is actually one huge molecule of DNA joined end-to-end in a circle. In addition, extrachromosomal DNA in circular plasmids, much smaller than the chromosome and non-essential to the cell, is commonly found, and a large number of bacteriophages which can infect *E. coli* have been described. All of these properties are valuable assets in an organism used for such diverse aspects of biological research.

Only in recent decades has *E. coli* been fully recognized as a pathogenic species in man and many domestic animals, previously having been regarded as a commensal inhabitant of the digestive tract. Various strains of the species are now known to be capable of causing a diverse range of diseases, from diarrhoea in both animals and man, mediated by cholera toxin-like and other toxins, through other forms of enteric disease resembling bacillary dysentery, and urinary tract infection, to neonatal meningitis. This diversity of disease syndromes reflects the diversity of determinants of pathogenic behaviour which the organism can express – although not all in the same strain! These include: adhesions, such as surface fimbriae which allow the organism to adhere to mucosal surfaces of the alimentary or urinary tract; a variety of toxins; a haemolysin; and polysaccharide capsules and lipopolysaccharides which enable it to resist phagocytosis and complement-mediated lysis in invasive diseases. It can express a wide variety of strain-specific surface antigens, of which the best-characterized can be serotyped by the use of specific antibody reagents in agglutination tests, into O- ('somatic' lipopolysaccharide), K- (capsular, usually polysaccharide), and H- (flagellar) types. In addition, fimbriae are also variable between strains, but as yet few fimbrial serotypes have been widely recognized.

Fortunately, for the laboratory worker, many isolates of *E. coli* possess none of the above virulence attributes, and indeed some well-known laboratory strains such as K12 which have been maintained for many years *in vitro* have permanently lost much of their capacity to express some of these factors even if the corresponding genes are introduced back into them by genetic manipulation. Virulence in a bacterium is usually the net result of interaction of multiple factors of the organism with the host, and seldom is a single attribute such as expression of a toxin able in itself to confer full pathogenic properties on a laboratory-adapted strain. Hence, work with defined laboratory strains of *E. coli* is essentially quite safe, provided proper common-sense procedures are followed.

Shigella

Shigella strains differ from *E. coli* primarily in their failure to ferment lactose (although they may have a beta-galactosidase, lacking only permease for the uptake of the substrate). Hence *Shigella* colonies may be pale pink on MacConkey's agar, due to trace fermentation of lactose, but not the bright magenta of *E. coli*. Normally also, *Shigella* strains are non-motile. In other respects they are almost identical to *E. coli*, with so much homology in their genetic and chromosomal organization that rationally they might be regarded as the same species. *Shigella* strains are classically the cause of bacillary dysentery, which unlike cholera and most diarrhoea caused by *E. coli*, involves invasion and destruction of the intestinal epithelium, known to be mediated in part by plasmid-encoded virulence factors.

The close relationship and yet distinct difference between *Shigella* and *E. coli* highlights an interesting observation in bacterial classification, noticeable when a large number of strains are analysed in detail. Often groupings of apparently unrelated properties (e.g. lactose non-fermentation and tendency to invade intestinal epithelium) are found to be common among a subgroup of the overall collection of related organisms. Several such groupings are found within the species *E. coli*, and are described as 'clonal', i.e. perhaps derived from a common ancestor strain. Thus, despite the possibilities for genetic information to move quite freely between strains of the same bacterial species, a surprising degree of conservation and stability seems to remain in nature.

Salmonella

Salmonella strains are slightly less closely related than *Shigella* to *E. coli*. Again they do not ferment lactose, and usually also they do not have beta-galactosidase. Like *E. coli* they have been studied very intensively in the laboratory, particularly by molecular geneticists although the *E. coli* model is now far more widely used in fundamental studies in molecular biology. Genetic manipulation has made possible new methods of gene transfer and the construction of 'unnatural' new combinations of genetic information, so considerations of biological safety demand that the better characterized and less potentially pathogenic *E. coli* is used primarily as host organism in such work.

The genus *Salmonella* is defined by its similar properties to *E. coli*, as a small, Gram-negative facultatively anaerobic non-spore-forming rod, characteristically non-lactose-fermenting but motile unlike *Shigella*. Again, it is usually capable of growth on minimal media.

Salmonella is essentially a pathogenic rather than a commensal species, although it does often have a close relationship with host animals, without causing overt disease, known as the 'carrier state'. Possibly this may develop from a disease condition but as host immunity develops during the course of infection, a form of equilibrium between host and pathogen is established. The classic example of a *Salmonella* carrier state is the well-known case of 'Typhoid Mary', who carried *Salmonella typhi*, the cause of human typhoid. She was a public health hazard for several years in New York early this century, responsible for the initiation of several major outbreaks of typhoid. In addition to the typhoid organism known as *S. typhi*, most other so-called species of *Salmonella* are really simply serotypes, differentiated from one another by agglutination tests with specific antisera as described for *E. coli* above. The other well-known pathogenic 'species', involved primarily in food poisoning in man but significant pathogens for domestic animals, include *S. typhimurium, S. dublin* and *S. enteritidis*. These also may best be regarded as 'clones' of the species *Salmonella*. Again, many of them are 'carried' by domestic animals, particularly by poultry, hence the recent publicity given to concern with hygiene in the poultry industry. In fact, this problem has been known for many years to bacteriologists and is not entirely the new phenomenon presented by the media!

Other commonly used laboratory organisms

Bacillus

Like *E. coli*, the genus *Bacillus* is widely used as a tool for teaching and for basic microbiological research, as one of the prototype representatives of the Gram-positive eubacteria, and of particular interest to biologists because the formation of spores is a valuable model of developmental processes in prokaryotic cell biology. *Bacillus* species tend to produce rather large, extravagant colonies, often with characteristic branching projections from the colony edge; some isolates spread very vigorously on agar, causing considerable problems as contaminants. It is essentially a saprophytic genus, with one or two very important exceptions, including *Bacillus anthracis*, the causative agent of anthrax and one of the most unpleasant bacterial pathogens when it 'accidentally' infects man – the usual host being cattle. Cases still occur occasionally among leather handlers in this country, emphasizing the ability of the spores to survive in adverse conditions – the reason incidentally for the interest shown in this bacterium by those interested in the potential application of microorganisms as biological warfare agents. Another 'pathogenic' species is *B. thuringiensis*, which produces a spore coat protein toxic for insects upon ingestion, and the organism is thus a potential agent of biological pest control. The genus comprises large, Gram-positive, endospore-forming rods many of which are facultatively anaerobic. They are usually fermentative, or undergo aerobic respiration or denitrification. They are common in the environment, and able to grow in a wide variety of conditions, both nutritionally and physically, with some very interesting adaptations to particular niches. *Bacillus subtilis* is a typical member of the genus (although a little smaller than most), ubiquitous in the environment and with spores which can withstand prolonged boiling – hence it is a common contaminant in bacteriology if autoclaves are not working or used properly. It is able to grow on minimal media, optimally at temperatures of about 37°C. *Bacillus cereus*, a larger-celled species, has several distinctive characters: its temperature optimum is about 30°C; it often produces very rhizoid, swirling colonies; and it produces an unusual and poorly characterized toxin if allowed to grow in large numbers in cooked rice. Ingestion of such spoiled rice can lead to rapid onset of food poisoning.

Another interesting property of some members of the genus *Bacillus* is their thermostability in the vegetative state. *Bacillus stearothermophilus* is a well-known example, growing at an optimum temperature of about 60°C and slow to grow at less than about 40°C: a thermophile, specially adapted for growth in decaying organic matter such as in compost heaps where rapid biological oxidation of organic substrates can lead to build-up of heat and raised temperatures. The thermostability of the degradative enzymes of this organism is exploited in the formulation of biological washing powders.

Clostridium

Many *Clostridium* species are obligate anaerobes, although they are often reasonably tolerant of short-term exposure to oxygen and so have been better recognized and characterized than many of the more labile genera of strict anaerobes which have only been investigated thoroughly quite recently, as methods for their handling have become widely available. *Clostridium* species are again large, Gram-positive, endospore-forming rods, grouped as a genus primarily because they share these properties as well as being strict anaerobes. Metabolically they tend to be degradative, often with potent proteases and other extracellular enzymes which presumably help them to survive by the breakdown of macromolecular substrates to release nutrients in decaying organic matter. Recent investigation of their rRNA sequences suggests that they are in fact rather a diverse group phylogenetically. They are important because of their potential to harm man and other animals, not usually as true 'parasitic' pathogens but through the production of potent toxins. These include the tetanus toxin of *C. tetani*, the incredibly potent neurotoxin of *C. botulinum*, cause of botulism, and the necrotizing gas-gangrene toxins and others of species such as *C. welchii*, often significant in diseases of domestic animals. A number of haemolysins, some with toxic activity and phospholipase activity, have been thoroughly investigated. The study of these factors is a science in itself, too detailed to go into here but providing many fascinating areas of overlap between microbiology and biochemistry.

One recent development in the medical microbiology of *Clostridium* has been the recognition of *C. difficile*, an organism that is able to colonize the gut of man in large numbers after disturbance of the normal balance of microbial gut flora caused by antibiotic treatment. The colonizing organisms harm the host, again due to the production of toxins which in this case act upon the gut to produce an acute inflammatory dysentery-like condition known as pseudomembranous colitis.

Staphylococcus

Staphylococci are one of the main groups of Gram-positive cocci which have been thoroughly studied. They grow characteristically in bunches or clusters, which is one of the main features distinguishing them from the streptococci, which tend to grow in long chains of cells joined together or 'strings' – a useful mnemonic for the student encountering these confusing names is the alliterative *str*- relating *str*ep to *str*ing! Staphylococci are closely related to the streptococci, and to other fermentative Gram-positive bacteria of the lactic acid group, like the lactobacilli. Colonies are very characteristic: large (2–3 mm on rich media after overnight incubation at 37°C), very opaque, round, smooth and either golden yellow (*Staphylococcus aureus*) or white (other species). They are facultatively anaerobic and while capable of lactic acid fermentation, also have respiratory metabolism. The staphylococci have been widely investigated and well-characterized and have considerable medical importance, although they are not often encountered as highly virulent pathogens in normal individuals. They are subdivided on a rather arbitrary basis into the pathogenic *S. aureus*, and the normally harmless collection of species broadly grouped under the label '*S. albus*', which include the 'species' *S. epidermidis* and *S. saprophyticus*, among others. Similar structurally, and in some cultural characteristics, are the micrococci, but they are quite distantly related by metabolic and phylogenetic criteria. Among the staphylococci, *S. aureus* is perhaps best regarded as another example of a clonal grouping; the distinguishing properties include the characteristic golden colony colour, possession of the extracellular fibrinogen-coagulating enzyme coagulase active on citrated blood plasma, an active DNase, and a variety of other extracellular proteins including haemolysins and other enzymes.

Staphylococci have been important model organisms for the study of cell-wall polymers, particularly peptidoglycan, and in the same context a model for investigation of the mechaniam of action of lysozyme and penicillin: the medical context of their isolation and study has been a spur for basic study. Another aspect of their role in medical microbiology is the frequent occurrence of antibiotic resistance mechanisms in these organisms; early in the history of the clinical use of penicillin, resistant staphylococci emerged. Penicillin-resistant staphylococci became common during the 1950s and 1960s, especially in hospitals, and while they occasionally caused severe disease in patients who were usually sick for other reasons, they have been a particular problem in hospitals as a reservoir of multiple antibiotic-resistance plasmids (see below – *Pseudomonas*). The problem persists today, with resistance to methicillin, virtually a last-resort antibiotic for some strains, now not uncommon among staphylococci.

Streptococcus

Streptococci comprise another larger and diverse genus of Gram-positive cocci, many of which have a tendency to grow in chains. They can be considered in several subgroups: the classical pathogen of lobar pneumonia, *Streptococcus pneumoniae*, also known as the pneumococcus or diplococcus, due to its tendency to grow in pairs rather than long chains; the beta-haemolytic group A streptococci, with the species name *S. pyogenes*; and a large number of other so-called 'species', differentiated largely by their serological group, type of haemolysis on blood agar and habitat. Some streptococci are pathogenic, while others are commensals. They are facultative anaerobes and are catalase-negative, and generally grow as very small colonies on agar. They are moderately fastidious organisms which grow best on rich media, and are not particularly robust. Upon growth on blood agar, the usual medium for their isolation, characteristic differences in their effects are seen: beta-haemolysis is a zone of complete clearing of the blood agar around the colony, while alpha-haemolysis is an incomplete clearing, leaving a greenish, translucent coloration. The former is associated with the pyogenic streptococci which may cause streptococcal sore throat, the latter mainly with avirulent, commensal strains, although serogroup is an important additional criterion for distinguishing strains and their pathogenic potential; there are a number of disease conditions associated with alpha-haemolytic streptococci of certain groups. Serogrouping is based on carbohydrate antigens in the cell wall, which can be extracted with hydrochloric acid and characterized by the 'Lancefield' typing scheme. Co-agglutination or latex agglutination tests are used for routine serogrouping.

Streptococci are important members of the normal microbial flora of the mucosal surfaces. They form probably the largest group in the mouth, where they colonize buccal and tongue surfaces by adhesion to epithelial cells, as well as the teeth where *S. mutans* is believed to play a crucial role in the establishment of dental plaque through its synthesis of insoluble, sticky glucans from sucrose, as well as participating in the build-up of acid through lactic acid fermentation. The latter is a significant feature of streptococcal metabolism – the organisms are important in the fermentation of milk to form yoghurt, for example. At the same time they are often, not surprisingly, acid-stable, and therefore survive the stomach acidity quite well and are common in the gut flora.

Pseudomonas

Pseudomonas is a very large and heterogeneous genus, into which a variety of organisms difficult to group with anything else have probably been 'lumped': phylogenetic analysis suggests widely diverse origins of some members of the genus. Nevertheless, well-characterized species such as *P. aeruginosa* and *P. fluorescens* are clearly related and useful subjects for laboratory study. They are aerobic, Gram-negative motile rods with polar flagella, with an oxidative, non-fermentative metabolism and are oxidase-positive. The above species produce characteristic pigments on nutrient agar or MacConkey plates: pyocyanin, which is blue (not produced by *P. fluorescens*), and fluorescein, a yellowish green pigment. Most organisms of the genus are nutritionally very non-exacting, able to utilize a very wide variety of carbon and nitrogen sources, to grow on minimal media, and to colonize a variety of environmental niches, particularly in moist or humid situations: they are basically saprophytes, which occasionally become problematic as potential pathogens. Their nutritional and metabolic versatility may enable them to colonize exotic habitats such as disinfectant solutions!

While not prominent primary pathogens in healthy people, pseudomonads, and especially *P. aeruginosa*, do cause serious problems in hospitals and in some patients, and in some circumstances in the pharmaceutical and toiletries industries. The reasons for this are twofold. First, the organisms characteristically have two attributes which make them difficult to eliminate either by antibiotic action in the host, or by the use of antiseptics and disinfectants. These are a highly impermeable and protective surface, especially the outer-membrane layer; and the frequent possession of genes encoding resistance mechanisms which counter the effects of antibiotics and other antibacterial agents. The outer membrane of Gram-negative bacteria has several components which may contribute to general resistance features: it is a permeability barrier, not only against low-molecular-weight solutes such as sugars and peptides, but also effective in excluding, for example, surfactant molecules such as the bile salts in the digestive tract which have potential antibacterial activity. The barrier against the latter is often the lipopolysaccharide component, and *Pseudomonas* species often have a highly effective lipopolysaccharide with high-molecular-weight, abundant polysaccharide side chains which keep surfactant molecules away from the vulnerable cytoplasmic membrane. Hydrophobic substances, including some antibiotics and other antibacterial substances, are also effectively excluded from the cell by this barrier. Essential water-soluble nutrients enter the periplasm by traversing the outer membrane through pores formed by specialized outer-membrane proteins known as porins. In *P. aeruginosa* these porins have unusually low permeability for many low-molecular-weight solutes including some more hydrophilic antibiotics, and hence the cells are often quite resistant intrinsically to a wide range of killing mechanisms. While this is not particularly important for people in normal health, who are not normally susceptible to infection by these organisms, those people who may have debilitating diseases or medical conditions and are hospitalized are often vulnerable through weakening of their immune and non-specific defence mechanisms.

The second resistance attribute which is common in *Pseudomonas* is the possession of resistance genes. In particular, *P. aeruginosa* is a notorious source of 'R-factor' multiple resistance plasmids, large plasmids which encode multiple resistance mechanisms active against antibiotics and other antibacterial substances including, for example, heavy metals. R factors are a particularly serious problem in medical microbiology because they not only encode multiple resistances, but they are able to transfer between a wide variety of bacterial hosts: an R-factor plasmid originating from a *Pseudomonas* species in a hospital might next be found in a strain of *Salmonella*, and indeed few effective antibiotics for the treatment of typhoid remain available to clinicians in certain countries, due to the spread of resistance mechanisms from other species through the 'promiscuity' of R-factor plasmids.

Other medically important and more fastidious species

Many familiar genus and species names will be encountered by the beginner in reading and learning, but these organisms are unlikely to be handled by the student for safety reasons or because they are difficult to grow. To inform the reader more widely, some of their practically important attributes will now be summarized.

A Gram-negative species of considerable importance as a pathogen is *Vibrio cholerae*; it is quite robust and will grow well on relatively crude media, probably reflecting its potential for survival and transmission from host to host in natural waters. This is a curved, non-sporing rod with polar flagella, aerobic, oxidase-positive although with fermentative capability, and tolerant of salt and alkalinity. The short, curved and highly motile rods with tapering ends are characteristic and have been likened to commas. The organism is a typical Gram-negative, with a serologically active lipopolysaccharide. The related *V. parahaemolyticus* is also a water-borne pathogen particularly prevalent in sea water and also capable of causing enteric disease.

Other non-fastidious Gram-negatives include *Pasteurella*, in which genus the agent of bubonic plague, now *Yersinia pestis*, was formerly classified: both are non-motile, small cocco-bacilli; *Pasteurella* is principally a pathogen of animals.

Several other Gram-negative pathogens are more or less fastidious, but most can be grown on blood agar or chocolate agar (the latter containing heat-coagulated blood). These include: *Brucella*, aerobic non-motile cocco-bacilli with oxidative metabolism; some species are highly pathogenic, primarily in cattle and sheep but also occasionally in man; brucellosis is a classically chronic, relapsing febrile condition (undulant fever), formerly prevalent among farm workers and veterinarians but now rare. While the risk of *Brucella* infection has been a major perceived hazard in recent years from the drinking of raw milk, the most likely such problems are now thought to arise from *Campylobacter* (see below). *Haemophilus* species, often present as a commensal in the human nasopharynx, are non-motile cocco-bacilli with a number of specific factors required for growth, with a preference for aerobic growth conditions; some strains are encapsulated, and can cause life-threatening meningitis. *Campylobacter*, a relatively newly named genus, includes both short and long (virtually spiral) curved motile rods with polar flagella, which grow well on chocolate blood agar and particularly thrive in an atmosphere enriched in CO_2 with reduced oxygen levels – microaerophiles. They include *C. jejuni*, formerly considered a *Vibrio*, a common cause of diarrhoea acquired from domestic animals, and the recently described *C. pylori*, suspected of a causative role in gastritis in man. *Bordetella*, the best-known species of which is the agent of whooping cough *B. pertussis*, are small aerobic cocco-bacilli, the latter non-motile, which have very exacting growth requirements. *Legionella*, which includes *L. pneumophila*, the agent of legionnaire's disease as well as some species less often implicated in disease, is slow to grow in culture and has a requirement for iron and cysteine and will not grow on blood agar. It is a small, Gram-negative, motile aerobic rod. Finally *Neisseria*, including both *N. gonorrhoeae*, agent of gonorrhoea, and *N. meningitidis*, a frequent commensal of the nasopharynx but also occasionally a serious pathogen able to cause severe meningitis, is also a relatively fastidious genus. These organisms require a rich basal medium with either lysed or chocolate-blood supplement, and have a strict requirement for raised CO_2 levels in the atmosphere, often provided by the use of a candle jar. They are very characteristically non-motile, Gram-negative diplococci, and can be identified with near-certainty on this basis in Gram-stained smears from appropriate clinical specimens, provided the clinical history supports a presumptive identification.

Significant obligate anaerobes include the very fastidious *Bacteroides* group, difficult to culture and to identify, and a prominent member of the large intestinal flora of many mammals. The organisms are small, Gram-negative, non-motile, non-sporing rods. This genus exemplifies the large group of fastidious, strict anaerobes which are easily killed by exposure to oxygen and have therefore received relatively little attention from bacteriologists due to the technical difficulties of working with them.

Gram-positives of importance as pathogens include *Corynebacterium diphtheriae*, the causative agent of diphtheria. This is a non-spore-forming, non-motile rod and grows best aerobically, on moderately rich medium supplemented with blood or serum. It is now very rarely isolated in developed countries since vaccines have virtually eliminated it from these populations. There are various other non-sporing, non-motile Gram-positive rods with similar morphology, of commensal origin, collectively known as the coryneforms or diphtheroids. A similar organism morphologically is *Listeria*, which is nutritionally unexacting, ubiquitous in the environment and will grow at a wide range of temperatures including those of ordinary refrigerators, hence their ability to contaminate stored dairy products. At high levels these organisms may be pathogenic for debilitated individuals or in pregnancy.

Mycobacterium is a very unusual bacterium of Gram-positive character, which is difficult and slow to culture and difficult to stain, due to a very thick and lipid-rich cell wall. After prolonged staining and exposure to acid, however, the stain which has penetrated remains fixed – they are 'acid fast'. The genus includes *Mycobacterium tuberculosis*, the tubercle bacillus, agent of tuberculosis, and *M. leprae*, the agent of leprosy: the latter is in fact non-cultivable on laboratory media. This group is very hazardous to handle in the laboratory. Other groups of slow-growing, Gram-positive type organisms which occasionally have a pathogenic role include *Actinomyces, Nocardia* and *Streptomyces*; they are slow to grow, often with a branching filamentous habit, and their handling is a specialized skill beyond the scope of this book. *Streptomyces* are very important organisms as sources of antibiotics, as well as numerous degradative enzymes which they may utilize as saprophytes.

Other bacteria which are outside the range of 'ordinary' laboratory organisms include the spirochaetes, spiral organisms of Gram-negative types which again are very difficult or impossible to culture in the laboratory and which include, for example, the syphilis agent *Treponema pallidum*. Two genera of obligate intracellular pathogens, most unusual among the bacteria, are *Chlamydia* and *Rickettsia*. Another group of very atypical bacteria is *Mycoplasma* in the group known as the Mollicutes, distinguished by their lack of rigid cell-wall structures. All of these require highly specialized facilities and skills for culture and experimentation.

YEASTS

The term 'yeast' is often used loosely to imply *Saccharomyces cerevisiae*, brewer's yeast, now a very important tool for basic research into eukaryotic molecular biology, rivalling in this role *E. coli*, its prokaryotic equivalent. Yeasts in fact comprise essentially the unicellular fungi, although some, notably *Candida albicans*, have alternative growth phases of mycelial habit – the phenomenon of dimorphism. For their routine culture, the most commonly used medium is Sabouraud's agar, an acidic (pH 5.4) peptone agar with added glucose (2 per cent w/v). The acidity of this medium makes it quite selective for yeasts, even when they are being cultured from environments rich in bacteria – many of which will not grow at such a low pH. Yeasts are significantly larger than bacteria – dimensions of perhaps $2–3 \times 4–5\,\mu m$ are not uncommon, and this difference makes cell suspensions considerably more opaque, relative to the numbers of cells present, than is the norm for bacteria. Yeast cells are also very robust; cell walls are thick, with several very stable structural polymers including chitin (poly-N-acetylglucosamine), beta-glucans, and mannans which are often covalently linked to proteins which in turn may be disulphide cross-linked. Yeast cells are therefore difficult to break open physically, but their walls can be degraded enzymically and by the action of reducing agents to produce protoplasts under osmotically stabilized conditions.

Further reading

Some of the following books are highly recommended as general background reading as well as for specific aspects of the material covered in this book. First the former are introduced with comments on their particular strengths, and this is followed by specific recommendations for further reading on the subject of each chapter.

General sources

General Microbiology, 5th edition, by R.Y. Stanier, J.L. Ingraham, M.L. Wheelis and P.R. Painter. 1987, Macmillan, London.

An excellent general textbook of microbiology with a long pedigree as a science-based introduction to the subject. This is an undergraduate level book, more advanced than is necessary for first-year undergraduate general biology students but an excellent reference source for those wishing to pursue the theoretical aspects of the subject further. There is however very little on practical methodology here.

General Microbiology, 6th edition, by H.G. Schlegel. 1986, Cambridge University Press, Cambridge.

A conveniently compact and affordable book for beginning undergraduates and possibly A-level students. Wide-ranging, with an emphasis on the biochemical and physiological aspects of the subject. Again, little detail of methodology.

Topley and Wilson's Principles of Bacteriology, Virology and Immunity, 7th edition. Volume 1, ***General Microbiology and Immunity***, edited by G.S. Wilson and H.M. Dick, 1983; Volume 2, ***Systematic Bacteriology***, edited by M.T. Parker, 1983; Volume 3, ***Bacterial Diseases***, edited by G.R. Smith, 1984. Edward Arnold, London.

Essentially a reference textbook, too expensive for most individuals to buy but a good compilation of contributions by experts in their field with a strong emphasis on the medical aspects of microbiology. All chapters are extensively referenced, both to important historical sources and to seminal recent publications. Useful reviews of methodology and laboratory practice are included in Volume 1.

Microbiology, 3rd edition, by B.D. Davis, R. Dulbecco, H.N. Eisen, H.S. Ginsberg and others. 1980, Harper and Row, Philadelphia.

Another classical textbook, again with a medical slant but sufficiently broad to give useful coverage in wider aspects of the subject. A large and relatively expensive volume, but good value although primarily theoretical rather than practical.

Practical Medical Microbiology, edited by J.G. Collee, J.P. Duguid, A.G. Fraser and B.P. Marmion. 1989, Churchill Livingstone, Edinburgh.

A comprehensive practical guide to medical bacteriology, with very useful detailed information on practical methodology and media and laboratory procedures. Although medical in context, much of the practical advice has a wider context. The book is however large (900 pp) and perhaps too expensive for use by undergraduates other than medical students for whom it is an excellent reference work.

Collins and Lyne's Microbiological Methods, 6th edition, edited by C.H. Collins, P.M. Lyne and J.M. Grange. 1989, Butterworths, London.

Again set in a medical context, this book nevertheless contains much useful advice on the organization and day-to-day use of a bacteriology laboratory. While rather detailed for elementary level teaching, it is a useful reference source and is particularly strong on safety aspects.

Chapter 2

Collins and Lyne's Microbiological Methods (see above).

Equipment needs are thoroughly described. Safety and the design and use of safety cabinets are also covered in detail. The fifth edition (1984), but not the sixth, has a comprehensive list of suppliers, many of them still accurate and relevant.

Anaerobic Bacteria, by K.T. Holland, J.S. Knapp and J.G. Shoesmith. 1987, Blackie, Glasgow.

An outline of methodology is given although there is little practical detail, and further references are listed.

Suppliers' catalogues

These often give very comprehensive listings of equipment which may be especially valuable in alerting laboratory workers to new equipment. They are normally available free to bona-fide institutional purchasers. Larger general laboratory catalogues are provided by, for example Gallenkamp, Orme Scientific, British Drug Houses, Scotlab (available generally only in Scotland) and Fisons. Trade journals such as ***Laboratory Equipment Digest*** (Morgan Grampian [Process Press] Ltd, 30 Calderwood Street, Woolwich, London SE18 6QH) also feature new developments and can be obtained free by genuine customers. In biotechnology and molecular biology commercial activity is well organized and financed at present, due to the booming supply industry for equipment and consumables. Prominent journals including ***Nature*** contain many advertisements by currently active suppliers.

Categorisation of Pathogens According to Hazard and Categories of Containment, Advisory Committee on Dangerous Pathogens. 1984, HMSO, London.

This booklet gives details of requirements for containment of pathogens in the laboratory, including the design and use of laboratories. It should be supplemented by the 'Howie' ***Code of Practice for the Prevention of Infection in Clinical Laboratories and Post-mortem Rooms*** (1978, DHSS; HMSO, London) which gives additional detailed advice.

Laboratory-acquired Infections, by C.H. Collins. 1983, Butterworths, London.

This includes comprehensive guidance on the design and siting of laboratories as well as details of the safety aspects of equipment.

Chapter 3

Light Microscopy in Biology: A Practical Approach, edited by A.J. Lacey. 1989, IRL Press, Oxford.

One of the well-known *Practical Approach* series, many of which, including this volume, are excellent value. There is detailed instruction and information on the use of the light microscope in most common applications in biology.

Chapter 4

Practical Medical Microbiology and ***Collins and Lyne's Microbiological Methods*** (see above).

These give full details of a wide range of media and their formulation. Also very useful reference sources are the manuals of media formulations provided by manufacturers, especially Difco and Oxoid.

Chapter 5

Collins and Lyne's Microbiological Methods (see above).

This gives additional details of sterilization and disinfection. Most good textbooks give some coverage of the theoretical aspects of bacterial death, although practical details are often omitted.

Theory and Practice of Experimental Bacteriology, 2nd edition, by G.G. Meynell and E. Meynell. 1970, Cambridge University Press, Cambridge.

This classic book includes both a thorough theoretical and quantitative treatment of all aspects of bacteriological practice including sterilization and disinfection, and practical details of methodology. Much of the material remains relevant although methods and materials are changing. Unfortunately the book has not been updated with a further edition.

Chapter 6

Laboratory-acquired Infections and ***Categorisation of Pathogens according to Hazard and Conditions of Containment*** (see above).

These both contain invaluable advice on all aspects of laboratory safety.

Pathogenesis of Infectious Disease, 3rd edition, by C.A. Mims. 1987, Academic Press, London.

A useful summary of pathogenic mechanisms and the microbial factors involved, which provides a good working knowledge of microbial pathogenicity.

Chapter 7

General Microbiology, by Stanier *et al.* (see above).

Theoretical aspects of microbial growth are adequately covered.

Biochemistry of Bacterial Growth, 3rd edition, edited by J. Mandelstam, K. McQuillen and I.W. Dawes. 1982, Blackwell, Oxford.

A detailed theoretical treatment of all aspects of microbial growth, beyond the needs of many undergraduate students but nevertheless a very useful reference source. The book is not practically oriented.

Chapter 8

See above references for Chapter 4.

Chapter 9

General Microbiology, by Stanier *et al.* (see above).
Theoretical aspects of quantitation of microbial growth are dealt with adequately.

Theory and Practice in Experimental Bacteriology by Meynell and Meynell (see above).
For greater theoretical and practical detail, this remains the most useful source for many applications.

Chapter 10

Bacterial Cell Surface Techniques, by I.C. Hancock and I.R. Poxton. 1988, John Wiley, Chichester.
This book includes useful information on processing bacterial cells, although little practical detail on harvesting. Its greatest value is in relation to subcellular fractionation (see below).

Centrifugation: A Practical Approach, 2nd edition, edited by D. Rickwood. 1984, IRL Press, Oxford.
Another very useful *Practical Approach* volume, one of the earliest produced. It contains a wealth of detailed information on centrifugation methodology as well as theoretical background.

Chapter 11

Animal Cell Culture: A Practical Approach, edited by R.I. Freshney. 1986, IRL Press, Oxford.
Some of the later volumes in the *Practical Approach* series, including this one, contain some advanced material which relates to quite highly specialized applications of the technology concerned. Nevertheless, there is some useful, broadly applicable practical detail on cell culture in this volume.

There appear to be few straightforward, elementary descriptions of cell-culture techniques available at present.

Chapter 12

Bacterial Cell Surface Techniques, by I.C. Hancock and I.R. Poxton (see above).
This is an invaluable source of information on molecular analysis of the bacterial cell.

Gel Electrophoresis of Proteins: A Practical Approach, 2nd edition, edited by B.D. Hames and D. Rickwood. 1989, IRL Press, Oxford.
An essential and uniquely powerful technology is here covered in excellent detail. A must for any worker using these methods for the first time.

Gel Electrophoresis of Nucleic Acids: A Practical Approach, 2nd edition, edited by D. Rickwood and B. Hames. 1990, IRL Press, Oxford.
The same comments apply as above.

Monoclonal Antibodies: Principles and Practice, 2nd edition, by J.W. Goding. 1986, Academic Press, London.
This book is unusual in giving a full range of practical instruction as well as a thorough theoretical treatment, and is highly recommended for anyone commencing work with monoclonal antibody systems.

Molecular Cloning: A Laboratory Manual, by T. Maniatis, E.F. Fritsch and J. Sambrook. 1982, Cold Spring Harbor Laboratory, New York.

Numerous books have appeared recently which describe in considerable detail the techniques of genetic manipulation in microbial systems. This reflects the boom in application of this technology in numerous fields of the life sciences and biomedicine. *Maniatis* is a classic and remains an essential handbook, especially for the relative newcomer to this technology, since it gives a logically structured account of all necessary techniques, intelligible even to the non-microbiologist. A second edition by J. Sambrook, E.F. Fritsch and T. Maniatis, has now been published (1989) and builds successfully upon the first.

Basic Cloning Techniques: A Manual of Experimental Procedures, edited by R.H. Pritchard and I.B. Holland. 1985, Blackwell, Oxford.

Several specific aspects of genetic manipulation are covered in full practical detail, although this is not a comprehensive guide. A usgeful source of selected methodology.

Principles of Gene Manipulation, 4th edition, by R.W. Old and S.B. Primrose. 1989, Blackwell, Oxford.

Although not concerned at all with the practicalities of gene manipulation, this book is highly recommended for its clear explanations of the basic principles involved.

Suppliers' Literature

Many catalogues, particularly from cell-culture and molecular-biology supply companies, contain a wealth of useful information in an immediately applicable form. Instructional material provided with kits (e.g. for cloning or sequencing DNA) is often particularly useful as a source of detailed protocols, and may also give standard literature references for the commonly used methods.

Chapter 13

Topley and Wilson's Principles of Bacteriology, Virology and Immunity (see above).

A wide range of standard microbiology textbooks exists, and all include a systematic treatment of diverse genera of bacteria. Many have a medical slant, and among them this volume is excellent.

General Microbiology, by Stanier *et al.* (see above).

This gives a recent and constructive view of the problems facing microbial systematists as molecular studies of evolution are superimposed on classical bacteriology, and also covers groups of no medical importance such as, for example, the archaebacteria.

Equipment suppliers

The following are major equipment suppliers, selected for the range of equipment or materials supplied, reputation, and in most cases the usefulness of catalogues or other information supplied.

General laboratory suppliers

BDH Chemicals Ltd, Broome Road, Poole, Dorset BH12 4NN

Traditionally suppliers of chemicals, BDH have more recently emphasized their capacity to supply equipment and other consumables, and are emerging as innovative suppliers of a range of more specialized kits and materials for bioscience.

Fisons Scientific Apparatus, Bishop Meadow Road, Loughborough, Leics LE11 0RG

While again pre-eminent as suppliers of chemicals, Fisons have also been prominent for many years as suppliers of general laboratory equipment and materials and have published a large and comprehensive catalogue useful as a reference source. A subsidiary is the centrifuge company **MSE**, no longer a manufacturer of large and ultracentrifuges but with excellent microfuges and benchtop models.

Gallenkamp, Belton Road West, Loughborough, Leics LE11 0TR

Another general supplier with a large range of reliable equipment including shakers, water baths and incubators. Again a comprehensive catalogue is published.

Sigma Chemical Company Ltd, Fancy Road, Poole, Dorset BH17 7NH

A major American chemical company, internationally recognized as a standard supplier of chemicals but also moving into many areas of bioscience with specialized catalogues, for example in serology and tissue cultures. The *Sigma Chemical Catalogue* is widely available in an annual edition and is a useful reference source for chemicals. The company guarantees despatch by post the same day if orders are received early in the day.

Scotlab, Unit 15, Earn Avenue, Righead Industrial Estate, Bellshill ML4 3JQ, Scotland

This general laboratory supplier is prominent in Scotland where a comprehensive general catalogue is issued; an innovative approach to basic laboratory facilities and materials is earning the company a wider reputation and many of their ideas are worth considering.

Whatman Labsales Ltd, Unit 1, Coldred Road, Parkwood, Maidstone, Kent ME15 9XN

Traditionally suppliers of filter papers and chromatography materials, the company remains prominent in separation technology but has recently expanded in the area of general laboratory supplies with some innovative approaches.

Media

Becton Dickinson UK Ltd, Between Towns Road, Cowley, Oxford, OX4 3LY
An international company with a wide range of media and other equipment including disposable syringes. Particularly known for the 'Gas-Pak' range of supplies for anaerobic and microaerophilic culture.

Difco Laboratories, PO Box 14B, Central Avenue, East Molesey, Surrey, KT8 0SE
Probably the best-known supplier of media internationally. Many products are regarded as 'standards' for research applications which require careful adherence to standard procedures for reproducibility, for example in molecular biology. An extensive catalogue contains much useful information and advice.

Lab M Ltd, Topley House, PO Box 19, Bury, Lancs BL9 6AU
A UK manufacturer which produces a full range of good quality microbiological media, often at somewhat lower prices than the well-known multinationals.

Oxoid Ltd, Wade Road, Basingstoke, Hants RG24 0PW
A long-established company in bacteriological media with a reputation rivalling that of Difco, but with some diversification into areas such as diagnostic reagents.

Tissue culture supplies

Flow Laboratories Ltd, Woodcock Hill, Harefield Road, Rickmansworth, Herts WD3 1PQ
One of the major multinational suppliers of cell-culture media and plastics, with an excellent and comprehensive catalogue which is a mine of useful information about cell lines, growth conditions and procedures.

Gibco Europe Ltd, Unit 4, Cowley Mill Trading Estate, Longbridge Way, Uxbridge, Middlesex UB8 2YG
A rival and equal competitor in range and quality of products and information with Flow Ltd. In addition the merged **BRL** is a leading supplier of high-quality molecular biology supplies and equipment.

Disposable plastics

Elkay Laboratories Ltd, Unit 2, Crockford Lane, Basingstoke, Hants RG24 0NA
One of many specialized companies which have entered the market for disposable plastics, especially polypropylene micropipette tips and Eppendorf tubes. This is a boom industry as disposal has steadily become more economical than washing up, and the single-use product more reliable for exacting procedures in molecular biology.

Sterilin Ltd, Sterilin House, Clockhouse Lane, Feltham, Middlesex TW14 8QS
An established supplier of plastic petri dishes and other disposables, especially in polystyrene.

Molecular biology supplies and equipment

BCL, Boehringer Corporation (London) Ltd, Boehringer Mannheim House, Bell Lane, Lewes, Sussex BN7 1LG
A wide range of reagents for molecular and cell biology, including all standard enzymes for DNA cloning and manipulation. The molecular biology catalogue is a useful source of information on methodology.

Bio-Rad Laboratories Ltd, Caxton Way, Watford Business Park, Watford, Herts WD1 8RP

A comprehensive range of chromatographic and electrophoretic equipment and materials for most separation and analysis needs in the microbiology laboratory.

Pharmacia Ltd, Pharmacia LKB Biotechnology Division, Midsummer Boulevard, Central Milton Keynes, Northants MK9 3HP

The merging of Pharmacia and LKB has created a uniquely broad range of reagents and equipment of the highest quality, for use in all molecular aspects of microbiology.

Culture collections

For a modest fee, the following will provide ampoules of bacterial cultures, normally freeze-dried under vacuum and available for delivery by post.

National Collection of Type Cultures, Central Public Health Laboratory, 61 Colindale Avenue, London NW9 5HT

National Collections of Industrial and Marine Bacteria Ltd, Torry Research Station, PO Box 31, 135 Abbey Road, Aberdeen, AB9 8DG, Scotland

American Type Culture Collection, 12301 Parklawn Drive, Rockville, Maryland 20852–1776, USA

Index